HF FILTER DESIGN
AND
COMPUTER SIMULATION

HF FILTER DESIGN
AND
COMPUTER SIMULATION

by

Randall W. Rhea
Founder and President
Eagleware Corporation

McGraw-Hill, Inc.

New York   San Francisco   Washington, D.C.   Auckland
Bogotá   Caracas   Lisbon   London   Madrid   Mexico City
Milan   Montreal   New Delhi   San Juan   Singapore
Sydney   Tokyo   Toronto

Library of Congress Cataloging-in-Publication Data

Rhea, Randall W.
  HF filter design and computer simulation / by Randall W. Rhea.
      p.   cm.
  Originally published: Atlanta : Noble Pub., ©1994.
  Includes bibliographical references and index.
  ISBN 0-07-052055-0
    1. Electric filters—Design and construction.   2. Electric
  filters—Mathematical models.   3. Microwaves—Mathematical models.
  I. Title.
  TK7872.F5R44   1995
  621.3815'324—dc20   95-1811
                                                            CIP

1 2 3 4 5 6 7 8 9 0   DOC/DOC   9 0 0 9 8 7 6 5

ISBN 0-07-052055-0

First published 1994 by NOBLE PUBLISHING, Atlanta, Georgia.

Printed and bound by R. R. Donnelley & Sons Company.

McGraw-Hill books are available at special quantity discounts to use as premiums and sales promotions, or for use in corporate training programs. For more information, please write to the Director of Special Sales, McGraw-Hill, Inc., 11 West 19th Street, New York, NY 10011. Or contact your local bookstore.

Noble L. (Bill) Rhea

1912 - 1985

He was born, reared, married, started a business, nurtured a family and retired from one street in a small midwestern community. But he was as worldly a person as you will find. He never met a stranger nor did he accept the existence of a mile or a thousand. Above all, he lived life to the fullest and instinctively knew what most of us never learn: life for him was a journey and not a destination. We miss you Dad.

Contents

Preface

Over the last several decades, modern filter theory has been significantly embellished by many contributors. In Zverev's [1] words "This search for useful theories has led to some of the most elegant mathematics to be found in the practical arts." Excitement over this elegance is tainted by sophistication more suited for the filter mathematician than the engineer whose work is often less specialized. This book is directed to the engineer and not the mathematician. We do so in full reverence of the mathematicians who provided the tools to work with in the trenches.

For completeness, a review of classic material is included, some of which predates WW II. Of course most of the material is more recent and some is original. It is at times a strange mixture, but always directed at the practical application of the art to today's real-world problems.

Chapters 1 through 5 cover fundamental concepts. Although this book emphasizes microwave filters, the first few chapters cover lumped element concepts more heavily than distributed elements. This is for two reasons. First, even at several gigahertz, lumped elements are useful when size is important, when stopband performance is critical and for MMIC processes. Second, much of the lumped element theory, with suitable modification, is applicable to distributed filter development.

Many engineers now at the peak of their careers began with slide rules. Less than one floating point multiply per second is performed to about three digits of precision using a slide rule.

Within a single generation, desktop computers revolutionized the way we design. Today, economic desktop computers deliver well over one million floating point operations per second at significantly improved precision. Any modern treatment of filters must acknowledge this power, so indeed, this book integrates numeric techniques with more classic symbolic theory. This is appropriate for a treatment emphasizing practical issues. Pure mathematics fatally falters when standard values, parasitics, discontinuities and other practical issues are considered. Chapter 6 is a review of available computer-aided filter techniques. Both simulation (design evaluation, optimization, tuning, and statistical analysis) and synthesis (finding topologies and element values to meet specifications) are covered. Examples use commercial software tools from Eagleware Corporation, although many of the presented techniques are suitable for other programs as well.

Chapters 7 through 10 cover distributed lowpass, bandpass, highpass and bandstop filters, respectively. No one filter type is optimal for all applications. The key to practical filter development is selection of the correct type for a given application. This is especially true for the bandpass class where the fractional bandwidth causes extreme variation in required realization parameters. Therefore, the largest variety of filter types are found in Chapter 8, which covers bandpass filters.

Appendix A covers PWB manufacture from the viewpoint of the design engineer who must work with service bureaus who specialize in board manufacture. Software tools discussed in Chapter 6 automatically plot artwork and/or write standard computer files for board manufacture. For greatest effectiveness, the designer should understand the limitations and constraints of the manufacturing process. Direct PWB milling equipment which provides same-day prototyping is also considered.

I would like to thank Rob Lefebvre who wrote Chapters 9, Chapter 10, Appendix A and Section 6.11. I am also indebted to Lu Connerley who typed, formatted and proofed for months.

This is the second book she has helped with, and I seem to recall promising we wouldn't do another after the first. Ryan Rhea helped prepare figures. John Taylor, Wes Gifford, Amin Salkhi, Richard Bell and Eric West of T-Tech provided several prototype milled PWBs. Advance Reproductions Corporation, MPC, Inc., and Lehighton Electronics, Inc. provided many example etched PWBs. Their addresses are given in Appendix A. Iraj Robati with Scientific-Atlanta provided time and equipment for some of the measured data presented in this book.

The largest heros are those listed as references. Many devoted their lives to the field and published their work for followers to build upon, layer by layer. In my current position, I enjoy daily discussion with design engineers. I have come to realize that although each engineering problem is unique, I often make referrals to the work of a few masters. It is to those masters that acknowledgement is truly due.

Randall W. Rhea
Stone Mountain, GA
July 21, 1993

[1] A. Zverev, The Golden Anniversary of Electric Wave Filters, *IEEE Spectrum*, March 1966, p. 129-131.

1

Introduction

This chapter is included for the novice. It provides a brief historical perspective and a review of very basic analog, high-frequency, electronic filter terminology.

1.1 Historical Perspective

As the wireless era began, selectivity was provided by a single series or shunt resonator. Modern filters date back to 1915 when Wagner in Germany and Campbell in the United States working independently proposed the filter [1]. In 1923, Zobel [2] at Bell Laboratories published a method for filter design using simple mathematics. His approximate "image parameter" technique was the only practical filter design method used for decades.

Around 1940, Foster's earlier theories were extended by Darlington and Cauer to exactly synthesize networks to prescribed transfer functions. Due to a heavy computational burden, these methods remained primarily of academic interest until digital computers were used to synthesize lowpass prototypes, from which other filter structures were easily derived. These lowpass prototypes have been tabulated for many specific and useful transfer approximations named after the mathematicians credited with the development of the polynomials, such as Butterworth, Chebyshev, Bessel, Gaussian and others. Although in practical use since the 1950s, this method is referred to as modern filter theory because it is the most recent of this triad of techniques.

At this point, two schools of interest developed. One pursued the extension and refinement of filter mathematics. For example, even digital computer precision is generally unsuitable when direct synthesis of bandpass instead of lowpass filters is attempted [3]. The final result of this pursuit is the ability to synthesize filters to nearly arbitrary requirements of passband and stopband response specified either by filter masks or transfer function polynomials.

The second school pursued the many problems associated with application of the lowpass prototype to the development of lowpass, bandpass, highpass and bandstop filters for practical applications. Problems include component parasitics, value realizability, differing reactor and resonator technologies, transmission line discontinuities, tunability and other issues. This pursuit involves development of a range of transformations from the lowpass prototype to various filter structures, each of which are well suited for certain applications. The results are numerous filters which maximize performance and realizability if the filter type and application are properly matched. These topics are the focus of this book. Early chapters review concepts and consider the reactor and resonator building blocks. The style of later chapters is case study.

1.2   Lowpass

A lowpass transmission amplitude response is given on the upper left in Figure 1-1. Energy from the source at frequencies lower than the cutoff frequency is transmitted through the filter and delivered to the output termination (load) with minimal attenuation.

A cascade of alternating series inductors and shunt capacitors forms a lowpass filter. At low frequency the reactances of the series inductors become very small and the reactances of the shunt capacitors become very large. These components effectively vanish and the source is connected directly to the load. When the termination resistances are equal, the maximum available energy from the source is delivered to the load. A singly-terminated class of filters exists with a finite termination

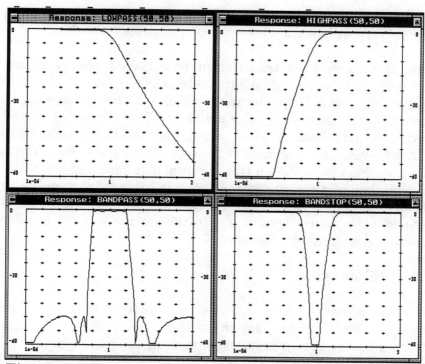

Figure 1-1 *Transmission amplitude response of lowpass (UL), highpass (UR), bandpass (LL) and bandstop filters (LR).*

resistance on one port and a zero or infinite termination resistance on the second port.

Lowpass element values may be chosen so that over a range of low frequencies, the element reactances cancel, or nearly cancel, and the impedance presented to the source as transformed through the network is nearly equal to the load. Again most of the energy available from the source is delivered to the load.

At higher frequencies, the series and shunt reactances become significant and impede energy transfer to the load. In a purely reactive network no energy is dissipated and, if it is not transmitted to the load, it is reflected back to the source. Energy not transmitted suffers attenuation (negative gain in decibel format). The ratio of reflected to incident energy in decibel format is return loss.

This transition from transmitted to reflected energy occurs suddenly only in an ideal filter. In a realizable filter, there exists a transition frequency range where increasing attenuation occurs with increasing frequency. The lowest stopband frequency has been reached when the rejection reaches the desired level. The steepness of the transition region (selectivity) is a function of the chosen transfer function approximation and the number of elements in the lowpass filter. The number of reactive elements in the all-pole lowpass prototype (all-pole is defined in a moment) is equal to the degree of the transfer function denominator (order).

1.3 Highpass

If each lowpass series inductor is replaced with a series capacitor, and each lowpass shunt capacitor is replaced with a shunt inductor, a highpass response such as that on the upper right in Figure 1-1 is achieved.

The highpass filter transfers energy to the load at frequencies higher than the cutoff frequency with minimal attenuation, and reflects an increasing fraction of the energy back to the source as the frequency is decreased below the cutoff frequency.

The transformation of lowpass series inductors to series capacitors and the lowpass shunt capacitors to shunt inductors is reasonably benign. In general, the realizability of inductor and capacitor (L-C) values in both lowpass and highpass filters is reasonably good; however, realization using element technologies other than L-C, such as transmission line (distributed), does pose some interesting problems.

1.4 Bandpass

A bandpass amplitude transmission response is given on the lower left in Figure 1-1. Energy is transferred to the load in a band of frequencies between the lower cutoff frequency, f_l, and the upper cutoff frequency, f_u. Transition and stopband regions occur both below and above the passband frequencies. The center frequency, f_o, is normally defined geometrically (f_o is equal to the square root of $f_u * f_l$).

One method of realizing a bandpass structure replaces each lowpass series inductor with a series L-C pair and replaces each lowpass shunt capacitor with a shunt, parallel resonant, L-C pair. This transformation results in a transfer function with double the degree of the original lowpass prototype. Shunning rigor in this book, we refer to the order of a bandpass structure as the order of the lowpass prototype from which it was derived.

The bandpass transformation is far from benign. For the lowpass it is only necessary to scale the lowpass prototype element values from the normalized values, at 1 ohm input termination and 1 radian cutoff frequency, to the desired values. For the bandpass, a new parameter is introduced, the fractional percentage bandwidth. Resulting bandpass element values are not only scaled by the termination impedance and center frequency, but they are modified by the fractional bandwidth parameter. This process has realizability implications, particularly for narrow bandwidth applications (the bandwidth between the lower and upper cutoff frequencies is small in relation to the center frequency). Realizability issues are addressed by utilization of alternative transformations. It is these alternative transformations which make bandpass filter design more involved and interesting than lowpass or highpass design.

The passband responses of the lowpass and highpass filters in Figure 1-1 are monotonic; the attenuation always increases with frequency as the corner frequency is approached from the passband. The bandpass response in Figure 1-1 has passband ripple. Because energy not transmitted is reflected, passband attenuation ripple results in non-monotonic return loss. Although generally undesirable, passband ripple is a necessary result of increased transition region steepness.

The lowpass and highpass responses in Figure 1-1 are also monotonic in the stopband; attenuation increases with increasing separation from the cutoff and reaches an infinite value only at infinite extremes of frequency (dc for the highpass and infinite frequency for the lowpass). This class of response is all-pole. It has only transmission poles and no transmission zeros at finite frequencies. The bandpass response in Figure 1-1c is not all-pole, but elliptic. It has infinite attenuation at finite frequencies in the stopbands.

1.5 Bandstop

A bandstop amplitude transmission response is given on the lower right in Figure 1-1. The bandstop transfers energy to the load in two frequency bands, one extending from dc to the lower bandstop cutoff and one extending from the upper bandstop cutoff to infinite frequency. The transition and stopband regions occur between the lower and upper cutoff frequency.

The lowpass prototype to bandstop transform suffers the same difficulties as the bandpass transform. Just as a bandpass filter offers improved selectivity over a single L-C resonator, the bandstop filter offers improved performance in relation to a "notch." Despite the obvious analogy, it is not uncommon for designers to attempt to improve notch performance by simply cascading notches instead of employing more effective bandstop filters. This is perhaps encouraged by the fact that bandstop applications are often intended to reject particular interfering signals, so the required stopband bandwidth is narrow which aggravates the difficulties with the true bandstop transformation.

1.6 All-Pass

To this point, we have been concerned with the amplitude transmission or reflection characteristics of filters. The ideal filter passes all energy in the desired bands and rejects all energy in the stopbands. The phase shift of transmitted energy in the ideal filter is zero, or at least linear with frequency (delayed only in time and otherwise undistorted). This is also not achieved in practice.

The rate of change of transmission phase with frequency is the group delay. Group delay is constant for linear transmission phase networks. Unfortunately, the group delay of selective, minimum-phase, networks is not flat, but tends to increase in magnitude (peak) near the corner frequencies. All passive ladder networks are minimum phase, and selectivity and flat group delay are mutually exclusive. Filter designs which begin with a controlled phase lowpass prototype, such as Bessel, result in excellent group delay flatness, but at the expense of selectivity.

A method of achieving both selectivity and flat delay consists of cascading a selective filter with a non-minimum phase network which has group delay properties which compensate the non-flat delay of the filter. A class of non-minimum phase networks with compensating delay characteristics but which do not disturb the amplitude characteristics of the cascade is referred to as all-pass.

1.7 Multiplexers

The above structures are two-port networks which selectively transmit or reflect energy. Couplers and splitters direct energy among multiple ports by dividing energy ideally without regard for frequency. A device which directs energy to ports based on the frequency band of the directed energy is referred to as a multiplexer. Because signal division occurs by frequency diversity, multiplexers offer the advantages of minimal loss in the desired bands and isolation across unwanted frequency bands.

A multiplexer typically has a common port and a number of frequency diversified ports. A multiplexer with a common port and two frequency diversified ports is referred to as a diplexer. A typical case includes a port driven by a lowpass filter and a port driven by highpass filter. Energy below a critical frequency is routed to the lowpass port and energy above a critical frequency is routed to the highpass port. Other specific terms such as triplexer and quadplexer are obvious.

When the 3 dB cutoff frequencies of the lowpass and highpass sections of such a diplexer are the same, the multiplexer is said to be contiguous. If the cutoff frequencies are spread by a guard band, the multiplexer is said to be non-contiguous. A multiplexer with three or more output ports may consist of both contiguous and non-contiguous bands.

The number of possible multiplexer combinations and variations is obviously endless. Fortunately, multiplexers are readily designed by designing individual filter sections and connecting them in parallel at the common port. This poses little difficulty provided a few points are considered. First, the terminal impedance behavior of each section should not interfere with the passband of any other section. This criteria is generally satisfied if series L-C resonators of bandpass multiplexers are connected together at the common port and fatally unsatisfied if the parallel shunt resonators are combined at the common port. Second, the useable bandwidth of elements must be sufficiently wide that parasitics do not invalidate the first criteria. Third, filter sections which are contiguous are designed as singly-terminated with the zero-impedance ports connected together to form the common port.

1.8 Additional References

Much of the original work on electric-wave filters is published in technical papers. Condensations of important works for the practicing engineer are found in two popular references, *Handbook of Filter Synthesis* by Zverev [4] and *Microwave Filters, Impedance-Matching Networks, and Coupling Structures* by Matthaei, Young and Jones [5]. Both of these timeless works have celebrated their silver anniversaries.

[1] A. Zverev, The Golden Anniversary of Electric Wave Filters, *IEEE Spectrum*, March 1966, p. 129.

[2] O. Zobel, *Theory and Design of Electric Wave Filters*, Bell System Technical Journal, January 1923

[3] H.J. Orchard and G.C. Temes, Filter Design Using Transformed Variables, *Trans. Circuit Theory*, December 1968, p. 90.

[4] A. Zverev, *Handbook of Filter Synthesis*, John Wiley and Sons, New York, 1967.

[5] G Matthaei, L. Young and E.M.T. Jones, *Microwave Filters, Impedance-Matching Networks, and Coupling Structures*, Artech House Books, Dedham, Massachusetts, 1967/1980.

2

Network Fundamentals

For this section, we assume that networks are linear and time invariant. Time invariant signifies that the network is constant with time. Linear signifies the output is a linear function of the input. Doubling the input driving function doubles the resultant output. The network may be uniquely defined by a set of linear equations relating port voltages and currents.

2.1 Voltage Transfer Functions

Consider the network in Figure 2-1 terminated at the generator with R_g, terminated at the load with R_l and driven from a voltage source E_g [1]. E_l is the voltage across the load.

The quantity E_{avail} is the voltage across the load when all of the available power from the generator is transferred to the load.

$$E_{avail} = \sqrt{\frac{R_l}{R_g}} \frac{E_g}{2} \tag{1}$$

For the case of a null network with $R_l = R_g$,

$$E_{avail} = \frac{E_g}{2} \tag{2}$$

since one-half of E_g is dropped across R_g and one-half is dropped across R_l. For the case of a non-null network, dividing both sides of equation (1) by E_l gives

$$\frac{E_{avail}}{E_l} = \sqrt{\frac{R_l}{R_g}} \frac{E_g}{2E_l} \tag{3}$$

We can then define the voltage transmission coefficient as the voltage across the load, E_l, divided by the maximum available voltage across the load E_{avail}, or

$$t = \frac{E_l}{E_{avail}} = \sqrt{\frac{R_g}{R_l}} \frac{2E_l}{E_g} \tag{4}$$

This voltage transmission coefficient is the "voltage gain" ratio.

2.2 Power Transfer Functions

The power insertion loss is defined as

$$\frac{P_{null}}{P_l} = \left(\frac{R_l}{R_l + R_g}\right)^2 \left|\frac{E_g}{E_l}\right|^2 \tag{5}$$

where the voltages and resistances are defined as before, P_{null} is the power delivered to the load with a null network and P_l is the power delivered to the load with a network present. Figure 2-2 depicts P_d as a function of R_l with a null network, $E_g=1.414$ volts and $R_g=1$ ohm. Notice the maximum power delivered to the load occurs with $R_l = 1$ ohm $= R_g$.

When R_l is not equal to R_g, a network such as an ideal transformer or a reactive matching network may reestablish maximum power transfer. When inserted, this passive network may therefore result in more power being delivered to the load than when absent. The embarrassment of power "gain" from a passive device is avoided by an alternative definition, the power transfer function

$$\frac{P_{avail}}{P_l} = \frac{R_l}{4R_g}\left|\frac{E_g}{E_l}\right|^2 = \frac{1}{t^2} \tag{6}$$

where

$$P_{avail} = \frac{|E_g|^2}{4R_g} \tag{7}$$

When $R_l=R_g$, these definitions are identical.

2.3 Scattering Parameters

The network depicted in Figure 2-1 may be uniquely described by a set of linear, time-invariant equations relating port voltages and currents. A number of two-port parameter sets including H, Y, Z, $ABCD$, S and others have been used for this purpose. Each have advantages and disadvantages for a given application. Carson [2] and Altman [3] consider network parameter sets indetail.

S-parameters have earned a prominent position in RF circuit design, analysis and measurement [4,5]. Other parameters such as Y, Z and H parameters, require open or short circuits on

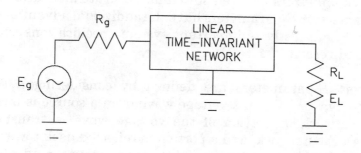

Figure 2-1 *Two-port network driven by a voltage source and terminated at both ports.*

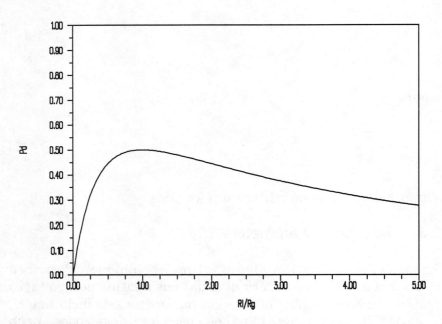

Figure 2-2 *Power delivered to the load versus the termination resistance ratio.*

ports during measurement. This poses serious practical difficulties for broadband high frequency measurement. Scattering parameters (S-parameters) are defined and measured with ports terminated in a reference impedance. Modern network analyzers are well suited for accurate measurement of S-parameters. S-parameters have the additional advantage that they relate directly to important system specifications such as gain and return loss.

Two-port S-parameters are defined by considering a set of voltage waves. When a voltage wave from a source is incident on a network, a portion of the voltage wave is transmitted through the network, and a portion is reflected back toward the source. Incident and reflected voltages waves may also be present at the output of the network. New variables are defined by dividing the voltage waves by the square root of the reference impedance. The square of the magnitude of these new variables may be viewed as traveling power waves.

$$|a_1|^2 = \text{incident power wave at the network input} \tag{8}$$

$$|b_1|^2 = \text{reflected power wave at the network input} \tag{9}$$

$$|a_2|^2 = \text{incident power wave at the network output} \tag{10}$$

$$|b_2|^2 = \text{reflected power wave at the network output} \tag{11}$$

These new variables and the network S-parameters are related by the expressions

$$b_1 = a_1 S_{11} + a_2 S_{12} \tag{12}$$

$$b_2 = a_1 S_{21} + a_2 S_{22} \tag{13}$$

$$S_{11} = \frac{b_1}{a_1}, a_2 = 0 \tag{14}$$

$$S_{12} = \frac{b_1}{a_2}, a_1 = 0 \tag{15}$$

$$S_{21} = \frac{b_2}{a_1}, a_2 = 0 \tag{16}$$

$$S_{22} = \frac{b_2}{a_2}, a_1 = 0 \tag{17}$$

Terminating the network with a load equal to the reference impedance forces $a_2 = 0$. Under these conditions

$$S_{11} = \frac{b_1}{a_1} \tag{18}$$

$$S_{21} = \frac{b_2}{a_1} \tag{19}$$

S_{11} is then the network input reflection coefficient and S_{21} is the forward voltage transmission coefficient t of the network. When the generator and load resistance are equal, the voltage transmission coefficient t defined earlier is equal to S_{21}. Terminating the network at the input with a load equal to the reference impedance and driving the network from the output port forces $a_1 = 0$. Under these conditions

$$S_{22} = \frac{b_2}{a_2} \tag{20}$$

$$S_{12} = \frac{b_1}{a_2} \tag{21}$$

S_{22} is then the output reflection coefficient and S_{12} is the reverse transmission coefficient of the network.

The S-parameter coefficients defined above are linear ratios. The S-parameters also may be expressed as a decibel ratio.

Because S-parameters are voltage ratios, the two forms are related by the simple expressions

$$|S_{11}| = \textit{input reflection gain } (dB) = 20 \log |S_{11}| \qquad (22)$$

$$|S_{22}| = \textit{output reflection gain } (dB) = 20 \log |S_{22}| \qquad (23)$$

$$|S_{21}| = \textit{forward gain } (dB) = 20 \log |S_{21}| \qquad (24)$$

$$|S_{12}| = \textit{reverse gain } (dB) = 20 \log |S_{12}| \qquad (25)$$

To avoid confusion, the linear form of S_{11} and S_{22} is often referred to as the reflection coefficient and the decibel form is referred to as the return loss. The decibel form of S_{21} and S_{12} are often simply referred to as the forward and reverse gain. With equal generator and load resistances, S_{21} and S_{12} are equal to the power insertion gain defined earlier.

The reflection coefficients magnitudes, $|S_{11}|$ and $|S_{22}|$, are less than 1 for passive networks with positive resistance. Therefore, the decibel input and output reflection gains, $|S_{11}|$ and $|S_{22}|$, are negative numbers. Throughout this book, S_{11} and S_{22} are referred to as return losses, in agreement with standard industry convention. Therefore, the expressions above relating coefficients and the decibel forms should be negated for S_{11} and S_{22}.

Input $VSWR$ and S_{11} are related by

$$VSWR = \frac{1 + |S_{11}|}{1 - |S_{11}|} \qquad (26)$$

The output $VSWR$ is related to S_{22} by an analogous equation. Table 2-1 relates various values of reflection coefficient, return loss, and $VSWR$.

The complex input impedance is related to the input reflection coefficients by the expression

$$Z_{input} = Z_o \frac{1+S_{11}}{1-S_{11}} \tag{27}$$

The output impedance is defined by an analogous equation using S_{22}.

2.4 The Smith Chart

In 1939, Philip H. Smith published an article describing a circular chart useful for graphing and solving problems associated with transmission systems [5]. Although the characteristics of transmission systems are defined by simple equations, prior to the advent of scientific calculators and computers, evaluation of these equations was best accomplished using graphical techniques. The Smith chart gained wide acceptance during an important developmental period of the microwave industry. The chart has been applied to the solution of a wide variety of transmission system problems, many of which are described in a book by Philip Smith [6].

The design of broadband transmission systems using the Smith chart involves graphic constructions on the chart repeated for selected frequencies throughout the range of interest. Although a vast improvement over the use of a slide rule, the process is tedious except for single frequencies and useful primarily for training purposes. Modern interactive computer circuit simulation programs with high-speed tuning and optimization procedures are much more efficient. However, the Smith chart remains an important tool as an insightful display overlay for computer-generated data. A Smith chart is shown in Figure 2-3.

Table 2-1 *Radially Scaled Reflection Coefficient Parameters.*

S_{11} (dB)	C_{11}	VSWR		S_{11} (dB)	C_{11}	VSWR
40.0	0.010	1.020		6.02	0.500	3.000
30.0	0.032	1.065		5.00	0.562	3.570
25.0	0.056	1.119		4.44	0.600	3.997
20.0	0.100	1.222		4.00	0.631	4.419
18.0	0.126	1.288		3.01	0.707	5.829
16.0	0.158	1.377		2.92	0.714	6.005
15.0	0.178	1.433		2.00	0.794	8.724
14.0	0.200	1.499		1.94	0.800	8.992
13.0	0.224	1.577		1.74	0.818	10.02
12.0	0.251	1.671		1.00	0.891	17.39
10.5	0.299	1.851		0.915	0.900	19.00
10.0	0.316	1.925		0.869	0.905	20.00
9.54	0.333	2.000		0.446	0.950	39.00
9.00	0.355	2.100		0.175	0.980	99.00
8.00	0.398	2.323		0.0873	0.990	199.0
7.00	0.447	2.615				

The impedance Smith chart is a mapping of the impedance plane and the reflection coefficient. Therefore, the polar form of a reflection coefficient plotted on a Smith chart provides the corresponding impedance. All values on the chart are normalized to the reference impedance such as 50 ohms. The magnitude of the reflection coefficient is plotted as the distance from the center of the Smith chart. A perfect match plotted on a Smith chart is a vector of zero length (the reflection coefficient is zero) and is therefore located at the center of the chart which is $1 + j0$, or 50 ohms. The radius of the standard Smith chart is unity. Admittance Smith charts and compressed or expanded charts with other than unity radius at the circumference are available.

Purely resistive impedances map to the only straight line of the chart with zero ohms on the left and infinite resistance on the right. Pure reactance is on the circumference. The complete circles with centers on the real axis are constant normalized resistance circles. Arcs rising upwards are constant normalized inductive reactance and descending arcs are constant normalized

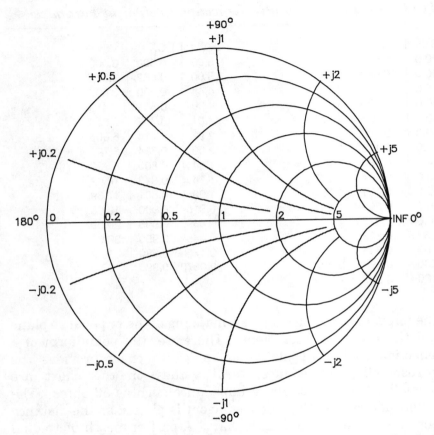

Figure 2-3 *Impedance Smith chart with unity reflection coefficient radius.*

capacitive reactance.

High impedances are located on the right portion of the chart, low impedances on the left portion, inductive reactance in the upper half, and capacitive reactance in the lower half. The angle of the reflection coefficient is measured with respect to the real axis, with zero degrees to the right of the center, 90 degrees straight up, and −90 degrees straight down. A vector of length 0.447 at 63.4 degrees extends to the intersection of the unity real circle and unity inductive reactance arc, $1 + j1$, or $50 + j50$ when denormalized.

The impedance of a load as viewed through a length of lossless transmission line as depicted on a Smith chart rotates in a clockwise direction with constant radius as the length of line or the frequency is increased. Transmission line loss causes the reflection coefficient to spiral inward.

2.5 Radially Scaled Parameters

The reflection coefficient, return loss, $VSWR$, and impedance of a network port are dependent parameters. A given impedance, whether specified as a reflection coefficient or return loss, plots at the same point on the Smith chart. The magnitude of the parameter is a function of the length of a vector from the chart center to the plot point. Therefore, these parameters are referred to as radially scaled parameters. For a lossless network, the transmission characteristics are also dependent on these radially scaled parameters. The length of this vector is the voltage reflection coefficient, ρ, and is essentially the reflection scattering parameter of that port. The complex reflection coefficient at a given port is related to the impedance by

$$\rho = \frac{Z - Z_0}{Z + Z_o} \tag{28}$$

where Z is the port impedance and Z_0 is the reference impedance. Then

$$RL_{dB} = -20 \log |\rho| \tag{29}$$

$$VSWR = \frac{1+|\rho|}{1-|\rho|} \tag{30}$$

$$L_A = -10\log(1-|\rho|^2) \tag{31}$$

Table 2-1 includes representative values relating these radially scaled parameters.

2.6 Modern Filter Theory

The ideal filter passes all desired passband frequencies with no attenuation and no phase shift, or at least linear phase, and totally rejects all stopband frequencies. The transition between pass and stopbands is sudden. This zonal filter is nonexistent. Modern filter theory begins with a finite-order polynomial transfer function to approximate the ideal response. Approximations are named after mathematicians credited with the development of the polynomial, such as Butterworth, Chebyshev and Bessel. In general, increasing polynomial order results in a more zonal (selective) response.

The filter is synthesized from the transfer function polynomial. A review of the required mathematics developed by a number of masterful contributors is given by Saal and Ulbrich [7].

2.7 Transfer Function

We begin by defining a voltage attenuation coefficient, H, which is the inverse of the previously defined voltage transmission coefficient

$$H = \frac{1}{t} \tag{32}$$

This voltage attenuation coefficient is variously referred to as the transfer function, voltage attenuation function, or the effective transmission factor. The attenuation of the network in decibels is

$$L_A = 20 \log |H| \qquad (33)$$

The transfer function may be expressed as the ratio of two polynomials in s where $s = \sigma + j\omega$. Therefore

$$H(s) = \frac{E(s)}{P(s)} \qquad (34)$$

The zeros of $H(s)$ are the roots of the numerator $E(s)$ and the poles of $H(s)$ are the roots of denominator $P(s)$. These roots may be depicted on a complex-frequency diagram as shown in Figure 2-4.

The horizontal axis of the complex-frequency diagram represents the real portion of roots and the vertical axis represents the imaginary portion of roots. Poles are indicated on the complex-frequency diagram as "x" and zeros are indicated as "o."

For realizable passive networks, the poles of $H(s)$ occur in the left half of the complex-frequency plane, or on the imaginary axis, while zeros may occur in either half. Poles and zeros occur in complex-conjugate pairs unless they lie on the real axis, in which case they may exist singly. For lossless ladder networks with no mutual inductors, $P(s)$ has only imaginary axis roots and is either purely even or purely odd.

2.8 Characteristic Function

Although practical filter networks utilize elements which include dissipative losses, for synthesis the network is assumed to include only reactive elements without loss. Therefore any power not transferred by the network to the load must be reflected back to the source. If we let $K(s)$ be a polynomial in s

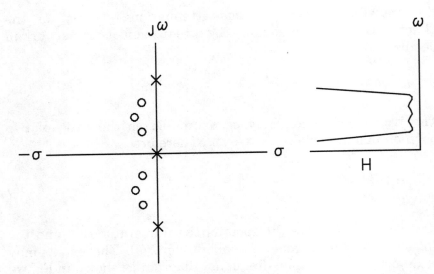

Figure 2-4 *Complex-frequency plane representation of a transfer function (left) and the corresponding magnitude versus frequency (right).*

for the ratio of the reflected voltage to the transferred voltage, we then have

$$|H(s)|^2 = 1 + |K(s)|^2 \qquad (35)$$

The above expression is referred to as the Feldtkeller equation and $K(s)$ is the characteristic function. The characteristic function $K(s)$ may be expressed as

$$K(s) = \frac{F(s)}{P(s)} \qquad (36)$$

Notice the denominators of the transfer function $H(s)$ and the characteristic function $K(s)$ are the same. Synthesis is possible once either $H(s)$ or $K(s)$ is known since the Feldtkeller equation relates these two functions.

2.9 Input Impedance

The filter reactive element values historically have been found from the input impedance of the network. The input impedance, reflection coefficient and network functions are related by

$$\rho_i = \left[\frac{Z_i - R_g}{Z_i + R_g}\right] = -\frac{K(s)}{H(s)} \tag{37}$$

Solving for the input impedance in terms of the network function polynomials we have

$$Z_i = R_g \left[\frac{E(s) - F(s)}{E(s) + F(s)}\right] \tag{38}$$

The actual element values are then found by a continued fraction expansion of Z_i.

2.10 Synthesis Example

Next, these concepts are applied to a lowpass filter with R_g equal to 1 ohm and ω_c equal to 1 radian. Consider the 3rd order transfer function

$$H(s) = s^3 + 2s^2 + 2s + 1 \tag{39}$$

Therefore

$$|H(s)|^2 = |H(s)| \, |H(-s)| = 1 + (-s^2)^3 = 1 - s^6 \tag{40}$$

In this case the denominator, $P(s)$, is simply unity and $E(s) = H(s)$. From the Feldtkeller equation

$$|K(s)|^2 = -s^6 \tag{41}$$

from which we infer $K(s) = s^3$. Again the denominator is unity so $F(s) = K(s)$.

Next these results are substituted in the expression for the input impedance of the terminated network to be synthesized and we have

$$Z_i = \left[\frac{s^3 + 2s^2 + 2s + 1 - s^3}{s^3 + 2s^2 + 2s + 1 + s^3} \right] = \left[\frac{2s^2 + 2s + 1}{2s^3 + 2s^2 + 2s + 1} \right] \tag{42}$$

This expression for the impedance of the terminated network is used to find the network element values. First we rationalize the numerator

$$Z_i = \frac{1}{\dfrac{2^3 + 2s^2 + 2s + 1}{2s^2 + 2s + 1}} \tag{43}$$

Next we continually divide the lower order polynomial into the higher order polynomial and invert the remainder. The final result is

$$Z_i = \cfrac{1}{s + \cfrac{1}{2s + \cfrac{1}{s + 1}}} \tag{44}$$

From this expansion, element values for the ladder network in Figure 2-5 are found as follows

$$Y_3 = 1s, \ C_3 = 1 \, farad \tag{45}$$

$$Z_2 = 2s, \ L_2 = 2 \, henries \tag{46}$$

$$Y_1 = 1s, \ C_1 = 1 \, farad \tag{47}$$

This network produces a lowpass response between 1 ohm generator and load resistances with 3.01 dB corner cutoff attenuation at a frequency of 1 radian per second.

2.11 Lowpass Prototype

This synthesized lowpass filter, normalized to 1 ohm and 1 radian cutoff, has series inductors in henries and shunt capacitors in farads. These values are referred to as prototype or g-values. This lowpass serves as a prototype for designing specific lowpass filters with other cutoff frequencies and termination resistances. These specific lowpass filters are easily found without resorting to the involved synthesis by simply scaling the g-values by the desired resistance and cutoff frequency. Designing highpass, bandpass, distributed, helical, distributed and other filters involves a transformation in addition to the scaling. Transformation and scaling are addressed after further discussion of transfer function approximations.

A diagram of this 3rd order lowpass prototype is given as the top schematic in Figure 2-6. The schematic on the right is an alternative form with a series inductor as the first element. The synthesized g-values are applicable to either form. Notice in the alternative form that $g(N+1)$ is inverted. In this case $g(N+1)$ is unity, but this is not always the case.

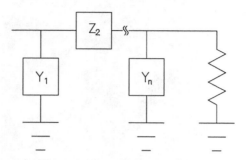

Figure 2-5 *Ladder network synthesized by continued fraction expansion with shunt susceptance and series impedance reactive elements and resistive terminations.*

2.12 Butterworth Approximation

The transfer function polynomial used in the previous synthesis example is the Butterworth approximation. The Butterworth approximation to a zonal filter is based on a maximally flat amplitude response constraint. Attenuation is low well within the passband and monotonically increases as the corner frequency(s) is approached. The transfer function poles fall on a circle in the complex-frequency plane. The performance and realizability properties of the Butterworth make it a natural choice for general purpose filter requirements.

Butterworth attenuation is given by the simple closed form expression

$$L_a = 10\log\left[1+\epsilon^2\left(\frac{f}{f_c}\right)^{2N}\right] \tag{48}$$

where N = the order, f = the desired frequency, f_c = the cutoff frequency and

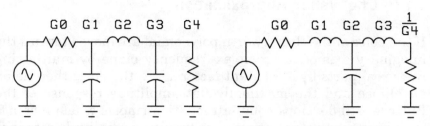

Figure 2-6 *Lowpass prototype structures. Notice the inversion of the output termination g-value in the alternate structure on the right.*

$$L_{Acutoff}=10\log(1+\epsilon^2) \tag{49}$$

With $\epsilon = 1$, $L_{Acutoff} = 3.01$ dB, which is the most popular definition of the cutoff attenuation for Butterworth filters.

In the lower portion of the passband, the attenuation is nil and therefore the return loss is excellent. Figure 2-7a gives the amplitude responses for Butterworth lowpass filters of order 5 and 7 up to two times the cutoff frequency. Given in Table 2-2 are Butterworth prototype values to 11th order. There are $N+2$ g-values in the prototype; N reactive values, $g(0)$, the normalized input termination resistance and $g(N+1)$, the normalized output termination resistance.

The above synthesis procedure is the basis of the Butterworth prototype g-values. With $\epsilon = 1$, Butterworth g-values are also given by the simple closed form expressions

$$g(n)=2\sin\left[\frac{(2n-1)\pi}{2N}\right],n=1,2,...,N \tag{50}$$

$$g(N+1)=1 \tag{51}$$

2.13 Chebyshev Approximation

If the poles of the Butterworth polynomial are moved toward the imaginary axis of the complex-frequency plane by multiplying their real parts by a constant factor $k_c < 1$, the poles then lie on an ellipse and the maximally flat amplitude response of the Butterworth develops equal-attenuation ripples which increase with increased pole shifting. The resulting amplitude response becomes more zonal and the selectivity increases. Even a small amount of ripple can significantly improve the selectivity. The approximation class is based on Chebyshev polynomials and is named accordingly. The shifting is related to the passband ripple, L_{Ar}, by

$$k_c = \tanh A \tag{52}$$

where

$$A = \frac{1}{N}\sinh^{-1}\left(\frac{1}{\epsilon}\right) \tag{53}$$

$$\epsilon = \sqrt{10^{L_{Ar}/10} - 1} \tag{54}$$

The amplitude responses for 5th and 7th order Chebyshev approximations are given as Figure 2-7b. The passband attenuation ripple L_{Ar} for these plots is 0.5 dB. For the moment the filter is assumed lossless, and since power not transmitted is reflected, the attenuation ripple results in passband return loss ripple. They are related by

$$RL_{dB} = -10\log \epsilon^2 \tag{55}$$

where ϵ is defined above.

Since increased ripple results in better selectivity, the Chebyshev approximation offers a compromise between passband ripple and selectivity. The Butterworth and 0 dB ripple Chebyshev are identical when the cutoff is suitably scaled.

The Chebyshev cutoff corner is defined by Williams and Zverev as 3.01 dB attenuation and by Matthaei as the ripple attenuation. The former is consistent with the Butterworth definition while the latter is consistent with a system performance viewpoint where return loss is an important specification.

With the cutoff attenuation defined as the ripple value, the Chebyshev amplitude response is given by

$$L_a = 10\log\left[1 + \epsilon^2 \cosh^2\left[N\cosh^{-1}\left(\frac{f}{f_c}\right)\right]\right] \tag{56}$$

where ϵ is defined above.

Lowpass prototype values may be found using the synthesis procedures previously outlined. Chebyshev g-values are also given by the closed form expressions [1, p. 99]

$$g(1) = 2\frac{a_1}{\gamma} \tag{57}$$

$$g(n) = 4\frac{a_{n-1}a_n}{b_{n-1}g_{n-1}} \tag{58}$$

where

$$\beta = \ln\left[\coth\left(\frac{R_{dB}}{17.37}\right)\right] \tag{59}$$

$$\gamma = \sinh\left(\frac{\beta}{2N}\right) \tag{60}$$

$$a_n = \sin\left[\frac{\pi(n-0.5)}{N}\right] \tag{61}$$

$$b_n = \gamma^2 + \sin^2\left(\frac{n\pi}{N}\right) \tag{62}$$

For N odd

$$g(N+1) = 1 \tag{63}$$

For N even

$$g(N+1) = \coth^2\left(\frac{\beta}{4}\right) \tag{64}$$

Lowpass prototype g-values for Chebyshev filters through 11th order are given in Tables 2-3 through 2-7.

2.14 Denormalization

The g-values are the inductance in henries and capacitance in farads for lowpass filters with the prescribed transfer function characteristics, a cutoff frequency of 1 radian, an input termination of $g(0)$ ohms (typically 1 ohm) and an output termination resistance of $g(N+1)$ ohms. Once these g-values have been found by synthesis, specific lowpass filters are designed by simply scaling the g-values by resistance and frequency factors as follows

$$L = \frac{gR}{\omega} \tag{65}$$

$$C = \frac{g}{R\omega} \tag{66}$$

where

$$\omega = 2\pi f \tag{67}$$

After scaling, the filter input termination resistance is R ohms, and the output termination impedance is R times $g(N+1)$. Since $g(0)$ equals $g(N+1)$ for Butterworth g-values, the input and output termination resistances are equal.

Notice that $G(0) = G(N+1)$ for all odd order Chebyshev and therefore the input and output termination resistances are equal. For all even order Chebyshev, $G(0) \neq G(N+1)$, and the output termination resistances are dissimilar. The number of inflections (zero slope occurrences) in the Chebyshev passband response is equal to the order and number of reactive elements in the Chebyshev prototype, as is evidenced in Figure 2-7b. Draw an example even order Chebyshev response. It becomes obvious that an even number of inflections dictates attenuation at dc. The dc attenuation magnitude is of course equal to the passband ripple. At dc the inductors and capacitors in a lowpass filter effectively vanish so attenuation must occur via mismatch. Therefore, the resistances terminating an even order true Chebyshev response cannot be equal. The terminating resistance ratio increases with increasing ripple.

2.15 Denormalization Example

Nearly all of the computational effort of lowpass filter design is stored in the lowpass prototype. All that remains to design a specific lowpass is scaling. Consider a 5th order, 2300 MHz cutoff Butterworth lowpass with the first element a series inductor. Using equation (65),

$$L_1 = \frac{0.61820 \times 50}{2\pi \times 2.3 \times 10^9} = 2.14\, nH \tag{68}$$

$$C_2 = \frac{1.618}{50 \times 2\pi \times 2.3 \times 10^9} = 2.24\, pF \tag{69}$$

$$L_3 = \frac{2.000 \times 50}{2\pi \times 2.3 \times 10^9} = 6.92\, nH \tag{70}$$

The Butterworth element values are symmetric so $C_4 = C_2$ and $L_5 = L_1$.

2.16 Phase and Delay

We have previously considered the amplitude responses of the filter transfer function. The transmission phase, ϕ, is the argument of the transfer function. While the transmission phase has long been important in such systems as video and radar, the expansion of data communications systems increases its significance.

A network with reactive elements must at least delay a transmitted signal in time. If the network passes all frequencies of the signal with equal amplitude attenuation and the transmission phase increases linearly with frequency, the signal is only delayed and level shifted, and it is not distorted.

The phase delay of a network is

$$t_p = \frac{\phi}{\omega} \tag{71}$$

and the group delay is

$$t_d = -\frac{\partial \phi}{\partial \omega} \tag{72}$$

The group delay is related to the time required for the envelope of a signal to transverse the network. It is also referred to as the envelope delay. Notice that if the transmission phase increases linearly with frequency, the group delay is constant. Therefore flat group delay is required if waveform distortion is to be avoided.

The lowpass prototype is a passive ladder network. The zeros must lie in the left half of the complex-frequency plane and such networks are referred to as minimum-phase. The phase and amplitude characteristics of minimum-phase networks are inseparably related via the Hilbert transform. Unfortunately, flat delay and good selectivity are mutually exclusive.

Shown in Figure 2-8 are the group delay responses of the filters with amplitude responses given in Figure 2-7. Notice the group delay asymptotically approaches a finite value as the frequency is decreased toward dc and tends to increase with frequency and peak in the vicinity of the cutoff. The exception is the Bessel transfer function which, however, possesses poor selectivity.

Because distortion is produced when different frequency components of a composite signal are unequally delayed, an important definition is differential delay. Differential group delay is the absolute difference in the group delay at two specified frequencies. Notice that more selective transfer functions exhibit greater differential delay within the passband.

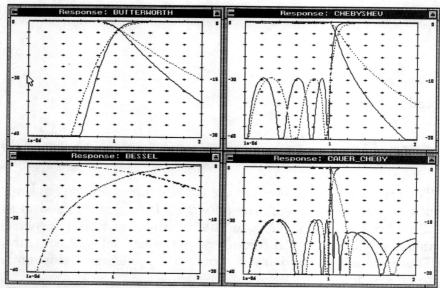

Figure 2-7 *Amplitude transmission and return loss for 5th-order (dashed) and 7th-order (solid) lowpass Butterworth (UL), Chebyshev (UR), Bessel (LL) and Cauer-Chebyshev (LR) filters.*

The Bessel transfer function has excellent group delay properties but poor selectivity.

2.17 Bessel Approximation

Just as a maximally flat amplitude response is approximated by Butterworth, a maximally flat group-delay response is approximated by the Bessel transfer function. The Gaussian transfer function is the basis of the Bessel. However, the Gaussian is not a closed formed polynomial but an infinite series. The Bessel, sometimes referred to as a Thompson, is a finite element approximation to Gaussian. Bessel prototype values are given in the prototype tables at the end of this chapter. There is significant amplitude attenuation well into the passband which results in poor return loss throughout much of the passband. The amplitude response of the Bessel is given approximately by

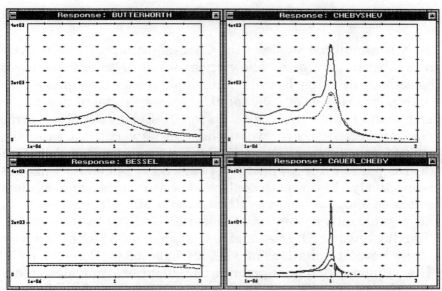

Figure 2-8 *Group delay responses for 5th-order (dashed) and 7th-order (solid) lowpass Butterworth (UL), Chebyshev (UR), Bessel (LL) and Cauer-Chebyshev (LR) filters.*

$$L_A \approx 3\left(\frac{f}{f_c}\right)^2 \tag{73}$$

Notice that unlike the Butterworth and Chebyshev, this approximate expression for the amplitude response is not a function of the filter order; the selectivity improves little with increasing order.

Notice the asymmetry in g-values for the Bessel prototype. With increasing order, the ratio of minimum and maximum element values becomes significant and contributes to realization difficulties.

2.18   Equiripple Phase-Error Approximation

Just as amplitude ripple in the Chebyshev improves selectivity over the Butterworth, allowing phase-error ripple improves selectivity over the Bessel. Equiripple phase-error prototypes with phase ripples of 0.05 and 0.5 degrees are tabulated at the end of the chapter for orders up to 10. Even with phase-error ripple, selectivity is far worse than Butterworth. At twice the cutoff frequency, the attenuation of Butterworth, Bessel and 0.5 degree phase-error lowpass filters are 30.1, 14.06 and 14.23 dB respectively. Notice both controlled phase filters are much less selective than the Butterworth and the equiripple phase-error is not significantly better than the Bessel. At five times the cutoff, the Butterworth, Bessel and 0.5 degree phase-error lowpass filters have attenuations of 69.9, 49.5 and 58.0 dB respectively. The equiripple phase-error approximation provides more rejection than the Bessel further into the stopband.

Other controlled phase approximations include transitional responses where the amplitude is similar to the Bessel over much of the passband (down to 6 or 12 dB) but which have improved selectivity above the transition region. Nevertheless, attempts at thwarting the selectivity versus flat group delay relationship of minimum-phase networks is largely futile.

2.19   All-Pass Networks

This fundamental limitation is circumvented by resorting to non-minimum-phase networks. These networks have transfer function zeros in the right half of the complex-frequency plane. They are realized as non-ladder networks with bridging paths or mutual inductances. Non-minimum-phase networks with a flat frequency response are all-pass. The magnitudes of the transfer function numerator and denominator must be related by a single constant at all frequencies. This requires that the right-plane zeros must be the mirror image of the left-plane poles. That is, the roots have the same imaginary-axis values but have real values of opposite sign.

Simultaneously selective and flat delay filters may be synthesized with embedded left-plane zeros [8,9]. A more popular approach is to cascade a conventional, selective, minimum-phase filter with one or more all-pass sections which compensate the differential delay of the filter. This is reviewed by Williams and Taylor [10].

A sufficient number of all-pass sections are cascaded to achieve a specified peak-to-peak differential group delay over a frequency band of interest which may include all or a portion of the passband. Given in Figure 2-9 on the left are the amplitude and group delay responses of a 5th order Butterworth lowpass filter with a cutoff of 6 MHz. On the right in Figure 2-9 are the amplitude and group delay responses for the lowpass cascaded with three all-pass sections designed to minimize the peak-to-peak differential group delay.

Each all-pass section requires several components whose values are dependent, making tuning difficult. A class of filters which are delay equalized to maximum flatness using a single all-pass section were developed by Rhea [11]. Selectivity for this filter class is near Butterworth. Lowpass prototype values are given in tables at the end of the chapter. In addition to the conventional all-pole lowpass g-values, included are the required 2nd order all-pass section normalized center frequency and Q values as defined in Williams. Figure 2-10 gives the amplitude and group delay responses of a 5th order singly-equalized filter designed using the table g-values. The rejection at 12 MHz is 6 dB less than the Butterworth, although it is far superior to a Bessel lowpass. Also, the delay is well equalized with only a single all-pass section and the absolute delay is far less than the equalized Butterworth.

Delay equalization is typically implemented at IF or baseband frequencies, and seldom at microwave frequencies. For that reason, the subject is not covered further here. The *Electronic Filter Design Handbook* [10] is an excellent treatment of the subject with coverage of both L-C and active all-pass networks.

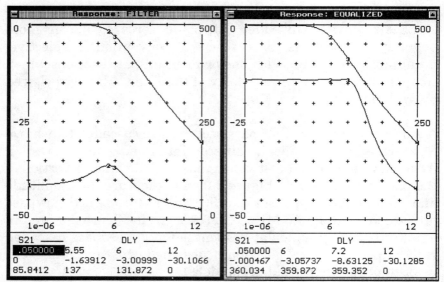

Figure 2-9 *Amplitude and group delay responses of a 6 MHz cutoff 5th order lowpass Butterworth filter before (left) and after (right) cascading with a 3 section all-pass group delay equalizer.*

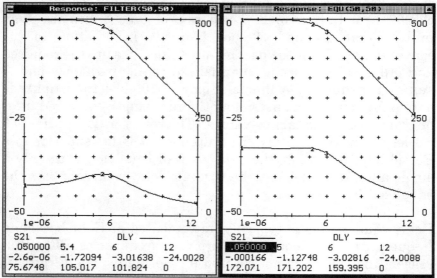

Figure 2-10 *Amplitude and group delay responses of a 6 MHz cutoff 5th-order singly-equalized lowpass filter.*

2.20 Elliptic Approximations

The lowpass transfer approximations previously discussed are all-pole; zeros of transmission occur only at infinite frequency. Series branches are purely inductive and shunt branches are purely capacitive. By incorporating a capacitor in parallel with series inductors or a series inductor with shunt capacitors, resonances are formed which cause a zero of transmission at finite frequencies. They provide increased transition region steepness at the expense of attenuation well into the stopband.

The class of filters with equiripple in the passband, L_{Ar} and equal minimum attenuation, A_{min}, in the stopband is referred to as Cauer-Chebyshev. The attenuation of Cauer-Chebyshev is given by

$$L_A = 10\log[1 + \epsilon^2 Z_n^2(\omega_n)] \tag{74}$$

where ω_n is the frequency normalized to the cutoff, ϵ is from the equation (54) and $Z_n(\omega_n)$ are elliptic functions of order N. The evaluation of $Z_n(\omega_n)$ and the attenuation expression above is involved and closed form solutions are unknown.

The lowpass prototype structure and the alternative form of the elliptic Cauer-Chebyshev are given in Figure 2-11. The amplitude and group delay responses of an example transfer function are given in Figures 2-7d and 2-8d, respectively. There are $(N-1)/2$ finite-frequency transmission zeros (and additional reactive elements) in odd order Cauer-Chebyshev prototypes and $(N-2)/2$ finite-frequency transmission zeros for even order. Equations (65) and (66) also scale elliptic prototype values.

Extensive tables of Cauer-Chebyshev prototype values to 7th order are given in Zverev [12]. Tables for Cauer-Chebyshev prototypes are lengthy because there are three independent continuous-value parameters, N, L_{Ar}, and A_{min}, which leads to a

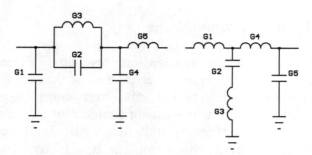

Figure 2-11 *Elliptic lowpass prototype structures.*

large number of combinations to tabulate. The Cauer-Chebyshev tables in Zverev include doubly and singly-terminated values and are categorized using the convention

$$CCn\rho\theta \qquad (75)$$

where CC signifies Cauer-Chebyshev, n is the order, ρ is the reflection coefficient in percent and θ is the modular angle. ρ is related to the passband ripple and other radially scaled parameters as given in equations (28-31) and Table 2-1. The modular angle is

$$\theta = \sin^{-1}\left(\frac{1}{\omega_s}\right) \qquad (76)$$

where ω_s is the cutoff-normalized lowest stopband frequency at which A_{min} occurs.

Amstutz [13] provides computer algorithms for Cauer-Chebyshev prototype values which describe distinctly different algorithms for input data of even and odd order. =FILTER= algorithms [14] remove this disadvantage. =FILTER= synthesizes both type B Cauer-Chebyshev filters with dissimilar termination resistances for even order and type C which approximate the response but with equal terminations.

Figure 2-12 compares the amplitude response of two 5th order Cauer-Chebyshev filters with L_{Ar} = 0.0436 dB to 5th and 7th order all-pole Chebyshev filter with the same passband ripple. On the upper left and upper right are 5th and 7th order Chebyshev, respectively. On the lower left and lower right are Cauer-Chebyshev with θ = 55 and 30 degrees respectively. A_{min} for these filters are approximately 26 and 55 respectively. The cutoff frequency in each case is unity. The responses are displayed through four times the cutoff.

The 5th order Chebyshev is included because the number of poles equals that of the Cauer-Chebyshev filters. When comparing economy, it should be recognized that 5th order

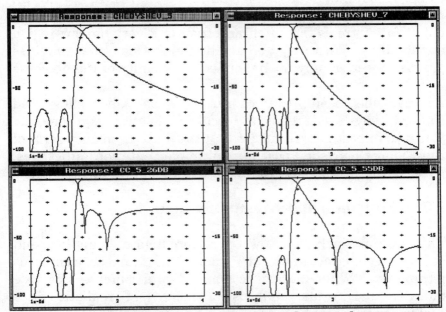

Figure 2-12 *Amplitude transmission and return loss responses of Chebyshev lowpass 5th (UL) and 7th (UR) order and 5th order Cauer-Chebyshev with 26 dB (LL) and 55 dB (LR) A_{min}.*

Cauer-Chebyshev lowpass have two additional components. Therefore, 7th order Chebyshev economy is equal to 5th order Cauer-Chebyshev economy. However, if inductors are considered more costly, this judgement is harsh because a 5th order Cauer-Chebyshev may have no more inductors than a 5th order Chebyshev.

Notice that in general, the Cauer-Chebyshev responses exhibit a steeper transition region, while further into the passband the all-pole responses exhibit greater rejection. When compared to the 5th order Chebyshev, with A_{min} = 26 dB, the Cauer-Chebyshev response provides greater rejection from the cutoff to approximately two times the cutoff. With A_{min} = 55 dB the Cauer-Chebyshev provides greater rejection from the cutoff to 3.8 times the cutoff.

When compared with the 7th order Chebyshev, the Cauer-Chebyshev is not as advantageous. With A_{min} = 26 dB, the Cauer-Chebyshev provides greater rejection only from the cutoff to 1.2 times the cutoff and then in a very narrow region at the notch. With A_{min} = 55 dB the Cauer-Chebyshev provides greater rejection only in the immediate vicinity of the notches.

As can be seen, the choice between Chebyshev and Cauer-Chebyshev depends on specific application requirements. It is also important to consider that the finite-frequency zeros of the Cauer-Chebyshev pose special implementation difficulties, particularly in some distributed microwave filter structures.

2.21　Bounding and Asymptotic Behavior

This section includes exact or approximate bounding and asymptotic characteristics for important filter performances parameters. While modern economic digital computers provide quick and accurate analysis of ideal and realistic filter performance throughout the entire frequency range of interest, asymptotic behavior provides important insight. This section covers behavior of ideal filters. We will consider asymptotic behavior relating to filter losses after we have considered component technologies and losses in a following chapter.

An approximate expression for attenuation well into the stopband of the lowpass is given by Cohn [13]. The series branch reactances are assumed to be high and the shunt branch reactances very low so that voltage dividers are simply cascaded.

$$L_A \approx 20\log\left[\omega_s^N \sum_{n=1}^{n=N} g(n)\right] - 10\log\left[\frac{4}{g(0)g(N+1)}\right] \tag{77}$$

where ω_s is the stopband frequency normalized to the 3 dB cutoff frequency. The summation of the reactive g-values is approximately $N = 1$, especially for the Butterworth transfer function. The accuracy of the above expression improves with increasing ω_s.

If $g(0) = g(N+1) = 1$ then

$$L_A \approx 20\log\left[\omega_s^N \sum_{n=1}^{n=N} g(n)\right] - 3.01 \tag{78}$$

For example, the summation of the reactive g-values is 6.494 for the 5th order 0.04365 dB ripple Chebyshev filter considered earlier. The ripple cutoff frequency is unity and the 3 dB cutoff frequency is approximately 1.185. At a stopband frequency of four, $\omega_s = 3.38$. Therefore the loss estimated by the above expression is

$$20\log(3.38^5 \times 6.494) - 3.01 = 66.1\,dB \tag{79}$$

while the actual attenuation is 63.6 dB.

When the termination resistances are similar, the group delay for the lowpass as $\omega \to 0$ is

$$t_{d0} \approx \frac{1}{2\omega_c} \sum_{n=1}^{n=N} g(n) \qquad (80)$$

Notice the expression for stopband loss used the unitless normalized stopband frequency while the expression for group delay uses the absolute cutoff frequency. With $\omega_c = 1$ MHz, both the low frequency group delay predicted by Cohn's expression and the actual group delay are 517 nS.

2.22 References

[1] G. Matthaei, L. Young and E.M.T. Jones, *Microwave Filters, Impedance-Matching Networks, and Coupling Structures*, Artech House Books, Norwood, Massachusetts, 1980, p. 36.

[2] R. Carson, *High-Frequency Amplifiers*, John Wiley & Sons, New York, 1962.

[3] J. Altman, *Microwave Circuits*, D. Van Nostrand, Princeton, New Jersey, 1964.

[4] Application Note 95, *S-Parameters - Circuit Analysis and Design*, Hewlett-Packard, Palo Alto, California, 1968.

[5] Application Note 154, *S-Parameter Design*, Hewlett-Packard, Palo Alto, California, 1972.

[6] Philip H. Smith, *Electronic Applications of the Smith Chart*, McGraw-Hill, New York, 1969.

[7] R. Saal and E. Ulbrich, *On the Design of Filters by Synthesis*, IRE Trans. Circuit Theory, December, 1958, pp. 284-317, 327.

[8] R. Lerner, Band-Pass Filters with Linear Phase, *Proc. IEEE*, March 1964, pp. 249-268.

[9] J.D. Rhodes and I.H. Zabalawi, Selective Linear Phase Filters Possessing a pair of j-Axis Transmission Zeros, *Circuit Theory and Applications*, Vol 10, 1982, pp. 251-263.

[10] A.B. Williams and F.J. Taylor, *Electronic Filter Design Handbook*, 2nd ed., McGraw-Hill, New York, 1988, pp. 7.1-7.28.

[11] R. Rhea, Singly-Equalized Filters, *RF Design*, October 1990, pp. 51-52 (insert).

[12] A. Zverev, *Handbook of Filter Synthesis*, John Wiley & Sons, New York, 1967, pp. 168-289.
[13] S. B. Cohn, Dissipation Loss in Multiple-Coupled-Resonator Filters, *Proc. IRE*, August 1959, pp. 1342-1348.

2.23 Prototype Tables

Digital computer programs which compute these values and automate many other filter design processes have made prototype g-value listings almost moot. However, for completeness, given here are some of the more important transfer function g-value listings. The listings are arranged by transfer function approximation. Each row is a set of g-values for a given order up to 10.

The generator (input) termination resistance is universally normalized to 1 ohm, that is $g(0) = 1$, so it is not included in the tables. The form of all-pole prototypes is shown in Figure 2-6 and the form of elliptic prototypes is shown in Figure 2-11.

Table 2-2 *Butterworth, A_a = 3.01 dB, G(0)=1.0.*

N	g(1)	g(2)	g(3)	g(4)	g(5)	g(6)	g(7)	g(8)	g(9)	g(10)	g(11)
2	1.4142	1.4142	1								
3	1.0000	2.0000	1.0000	1							
4	0.7654	1.8478	1.8478	0.7654	1						
5	0.6180	1.6180	2.0000	1.6180	0.6180	1					
6	0.5176	1.4142	1.9318	1.9318	1.4142	0.5176	1				
7	0.4450	1.2470	1.8019	2.0000	1.8019	1.2470	0.4450	1			
8	0.3902	1.1111	1.6629	1.9616	1.9616	1.6629	1.1111	0.3902	1		
9	0.3473	1.0000	1.5321	1.8794	2.0000	1.8794	1.5321	1.0000	0.3473	1	
10	0.3129	0.9080	1.4142	1.7820	1.9754	1.9754	1.7820	1.4142	0.9080	0.3129	1

Table 2-3 *Chebyshev, $A_a=L_{Ar}=0.01$ dB, R.L.=26.4 dB, G(0)=1.0.*

N	g(1)	g(2)	g(3)	g(4)	g(5)	g(6)	g(7)	g(8)	g(9)	g(10)	g(11)
2	0.4489	0.4078	0.9085								
3	0.6292	0.9703	0.6292	1							
4	0.7129	1.2004	1.3213	0.6476	0.9085						
5	0.7563	1.3049	1.5773	1.3049	0.7563	1					
6	0.7814	1.3600	1.6897	1.5350	1.4970	0.7098	0.9085				
7	0.7970	1.3924	1.7481	1.6331	1.7481	1.3924	0.7970	1			
8	0.8073	1.4131	1.7824	1.6833	1.8529	1.6193	1.5555	0.7334	0.9085		
9	0.8145	1.4271	1.8044	1.7125	1.9058	1.7125	1.8044	1.4271	0.8145	1	
10	0.8197	1.4370	1.8193	1.7311	1.9362	1.7590	1.9055	1.6528	1.5817	0.7446	0.9085

Table 2-4 *Chebyshev, $A_a=L_{Ar}=0.0432$ dB, R.L.=20 dB, G(0)=1.0*

N	g(1)	g(2)	g(3)	g(4)	g(5)	g(6)	g(7)	g(8)	g(9)	g(10)	g(11)
2	0.6648	0.5445	0.8190								
3	0.8516	1.1032	0.8516	1							
4	0.9314	1.2920	1.5775	0.7628	0.8190						
5	0.9714	1.3721	1.8014	1.3721	0.9714	1					
6	0.9940	1.4131	1.8933	1.5506	1.7253	0.8141	0.8190				
7	1.0080	1.4368	1.9398	1.6220	1.9398	1.4368	1.0080	1			
8	1.0171	1.4518	1.9667	1.6574	2.0237	1.6107	1.7726	0.8330	0.8190		
9	1.0235	1.4619	1.9837	1.6778	2.0649	1.6778	1.9837	1.4619	1.0235	1	
10	1.0281	1.4690	1.9952	1.6906	2.0882	1.7102	2.0642	1.6341	1.7936	0.8420	0.8190

Table 2-5 *Chebyshev, $A_a=L_{Ar}=0.10$ dB, R.L.=16.4 dB, G(0)=1.*

N	g(1)	g(2)	g(3)	g(4)	g(5)	g(6)	g(7)	g(8)	g(9)	g(10)	g(11)
2	0.8431	0.6220	0.7378								
3	1.0316	1.1474	1.0316	1							
4	1.1088	1.3062	1.7704	0.8181	0.7378						
5	1.1468	1.3712	1.9750	1.3712	1.1468	1					
6	1.1681	1.4040	2.0562	1.5171	1.9029	0.8618	0.7378				
7	1.1812	1.4228	2.0967	1.5734	2.0967	1.4228	1.1812	1			
8	1.1898	1.4346	2.1199	1.6010	2.1700	1.5641	1.9445	0.8778	0.7378		
9	1.1957	1.4426	2.1346	1.6167	2.2054	1.6167	2.1346	1.4426	1.1957	1	
10	1.2000	1.4482	2.1445	1.6266	2.2254	1.6419	2.2046	1.5822	1.9629	0.8853	0.7378

Table 2-6 Chebyshev, $A_a=L_{Ar}=0.20$ dB, R.L.=13.5 dB, $G(0)=1.0$.

Ng	g(1)	g(2)	g(3)	g(4)	g(5)	g(6)	g(7)	g(8)	g(9)	g(10)	g(11)
2	1.0379	0.6746	0.6499								
3	1.2276	1.1525	1.2276	1							
4	1.3029	1.2844	1.9762	0.8468	0.6499						
5	1.3395	1.3370	2.1661	1.3370	1.3395	1					
6	1.3598	1.3632	2.2395	1.4556	2.0974	0.8838	0.6499				
7	1.3723	1.3782	2.2757	1.5002	2.2757	1.3782	1.3723	1			
8	1.3804	1.3876	2.2964	1.5218	2.3414	1.4925	2.1349	0.8972	0.6499		
9	1.3861	1.3939	2.3094	1.5340	2.3728	1.5340	2.3094	1.3939	1.3861	1	
10	1.3901	1.3983	2.3181	1.5417	2.3905	1.5537	2.3720	1.5066	2.1514	0.9035	0.6499

Table 2-7 Chebyshev, $A_a=L_{Ar}=0.50$ dB, R.L.=9.6 dB, $G(0)=1.0$.

N	g(1)	g(2)	g(3)	g(4)	g(5)	g(6)	g(7)	g(8)	g(9)	g(10)	g(11)
2	1.4029	0.7071	0.5040								
3	1.5963	1.0967	1.5963	1							
4	1.6704	1.1926	2.3662	0.8419	0.5040						
5	1.7058	1.2296	2.5409	1.2296	1.7058	1					
6	1.7254	1.2478	2.6064	1.3136	2.4759	0.8696	0.5040				
7	1.7373	1.2582	2.6383	1.3443	2.6383	1.2582	1.7373	1			
8	1.7451	1.2647	2.6565	1.3590	2.6965	1.3389	2.5093	0.8795	0.5040		
9	1.7505	1.2690	2.6678	1.3673	2.7240	1.3673	2.6678	1.2690	1.7505	1	
10	1.7543	1.2722	2.6755	1.3725	2.7393	1.3806	2.7232	1.3484	2.5239	0.8842	0.5040

Table 2-8 Bessel, $A_a=3.01$ dB, $G(0)=1.0$.

N	g(1)	g(2)	g(3)	g(4)	g(5)	g(6)	g(7)	g(8)	g(9)	g(10)	g(11)
2	0.5755	2.1478	1								
3	0.3374	0.9705	2.2034	1							
4	0.2334	0.6725	1.0815	2.2404	1						
5	0.1743	0.5072	0.8040	1.1110	2.2582	1					
6	0.1365	0.4002	0.6392	0.8538	1.1126	2.2645	1				
7	0.1106	0.3259	0.5249	0.7020	0.8690	1.1052	2.2659	1			
8	0.0919	0.2719	0.4409	0.5936	0.7303	0.8695	1.0956	2.2656	1		
9	0.0780	0.2313	0.3770	0.5108	0.6306	0.7407	0.8639	1.0863	2.2649	1	
10	0.0672	0.1998	0.3270	0.4454	0.5528	0.6493	0.7420	0.8561	1.0781	2.2641	1

Table 2-9 *Equiripple Phase-Error 0.05°, A_a=3.01 dB, $G(0)$=1.0.*

N	g(1)	g(2)	g(3)	g(4)	g(5)	g(6)	g(7)	g(8)	g(9)	g(10)	g(11)
2	.6480	2.1085	1								
3	.4328	0.0427	2.2542	1							
4	.3363	0.7963	1.1428	2.2459	1						
5	.2751	0.6541	0.8892	1.1034	2.2873	1					
6	.2374	0.5662	0.7578	0.8760	1.1163	2.2448	1				
7	.2085	0.4999	0.6653	0.7521	0.8749	1.0671	2.2845	1			
8	.1891	0.4543	0.6031	0.6750	0.7590	0.8427	1.0901	2.2415	1		
9	.1718	0.4146	0.5498	0.6132	0.6774	0.7252	0.8450	1.0447	2.2834	1	
10	.1601	0.3867	0.5125	0.5702	0.6243	0.6557	0.7319	0.8178	1.0767	2.2387	1

Table 2-10 *Equiripple Phase-Error 0.5°, A_a=3.01 dB, $G(0)$=1.0.*

N	g(1)	g(2)	g(3)	g(4)	g(5)	g(6)	g(7)	g(8)	g(9)	g(10)	g(11)
2	.8245	1.9800	1								
3	.5534	1.0218	2.4250	1							
4	.4526	0.7967	1.2669	2.0504	1						
5	.3658	0.6768	0.9513	1.0113	2.4446	1					
6	.3313	0.5984	0.8390	0.7964	1.2734	2.0111	1				
7	.2876	0.5332	0.7142	0.6988	0.9219	0.9600	2.4404	1			
8	.2718	0.4999	0.6800	0.6312	0.8498	0.7447	1.3174	1.9626	1		
9	.2347	0.4493	0.5914	0.5747	0.7027	0.6552	0.8944	0.9255	2.4332	1	
10	.2359	0.4369	0.5887	0.5428	0.7034	0.5827	0.8720	0.6869	1.4317	1.8431	1

Table 2-11 *Singly-Equalized, A_a=3.01, $G(0)$=1.0, $G(N+1)$=1.0.*

N	g(1)	g(2)	g(3)	g(4)	g(5)	g(6)	g(7)	g(8)	g(9)	f_o	Q
2	1.4140	1.4140								3.8090	.2815
3	0.9222	2.0780	0.9222							1.1890	.5273
4	0.5917	1.8280	1.8280	0.5917						1.0640	.5400
5	0.4158	1.3810	2.1120	1.3810	0.4158					0.9707	.5525
6	0.4164	1.0940	1.9560	1.9560	1.0940	0.4164				0.9170	.5600
7	0.5199	1.0000	1.8190	1.9050	1.8190	1.0000	0.5199			0.8280	.5650
8	0.4804	0.9760	1.4120	1.9750	1.9750	1.4120	0.9760	.48040		0.8040	.5700
9	0.2742	0.8726	1.1000	1.9200	1.8420	1.9200	1.1000	.87260	.2742	0.7745	.5750

3

Reactors and Resonators

In a departure from the normal convention of relegating components to the back of the book, we take up the subject early because the realities of components cannot be divorced from the final assembly. The engineer who begins assembly design without consideration of more mundane component issues, and who assumes inductors are inductors or lines are lines, will come to believe that high-frequency design is black magic.

This chapter is not complete. Volumes have been written on these subjects. However, important issues are reviewed in a unified way to form a firm foundation for practical filter development which is studied later.

3.1   Inductance

Current flowing in a conductor produces a magnetic flux which encircles the current. When the conductor is arranged such that the flux encircles the conductor more effectively, such as by coiling the conductor, the flux linkage is increased. Inductance, L, is proportional to this flux linkage. Energy is stored in this magnetic flux. The stored energy is

$$Energy = \frac{1}{2}LI^2 \tag{1}$$

where I is the conductor current.

A change in current flow causes a change in the flux linkage. This flux change induces a voltage which attempts to resist the change in current. The inductor therefore has current inertia.

From a circuit viewpoint, the ideal inductor terminal impedance is

$$Z = j\omega L \tag{2}$$

The impedance is purely reactive, positive, and increases linearly with frequency.

3.2 Capacitance

An electric field is created when a potential difference is applied across conductors separated by an insulator (dielectric). Energy is stored in this electric field. The stored energy is

$$Energy = \frac{CE^2}{2} \tag{3}$$

where C is the capacitance and E is the potential difference. The capacitor attempts to retain a stored charge and maintain a constant potential difference. The terminal impedance is

$$Z = \frac{1}{j\omega C} = -j\frac{1}{\omega C} \tag{4}$$

The impedance is purely reactive, negative, and inversely proportional to frequency.

3.3 Unloaded Q

The ideal inductor and capacitor exhibit the above terminal circuit behavior and have no dissipated energy. With ideal elements, filter design would be pure mathematics and far simpler than it is in practice. Unfortunately, components exhibit loss and other parasitics. Loss occurs with electric fields in lossy dielectrics, with current flowing in lossy conductors and via radiation. Various component technologies have significantly different loss mechanisms and magnitudes. Just as importantly, the circuit configuration influences how a given component performs. For example, the midband loss in a bandpass filter is not only a function of component quality but also the design bandwidth. Circuit configuration effects are discussed in a later chapter.

Component Q, also referred to as unloaded Q, is defined as the ratio of the stored to dissipated energy in the element. Energy is stored in fields[1] and dissipated in resistance. For lumped elements, if the loss resistance is modeled as being in series with the reactance, X, the unloaded Q is

$$Q_u = \frac{X}{R_s} \tag{5}$$

From the above reactance expressions, Q_u for the inductor is

$$Q_u = \frac{\omega L}{R_s} \tag{6}$$

and for the capacitor is

$$Q_u = \frac{R_s}{\omega C} \tag{7}$$

If the loss resistance and reactance are modeled in parallel, the unloaded Q is

$$Q_u = \frac{R_p}{X} \tag{8}$$

which for the inductor and capacitor are, respectively,

[1]From a circuit viewpoint, reactance is a quantitative measure of fields.

$$Q_u = \frac{R_p}{\omega L} \tag{9}$$

Unloaded Q is a measure of component quality. The maximum available unloaded Q varies among component technologies. Finite unloaded Q results in filter passband insertion loss, heating in power applications, amplitude and delay response shape perturbation, return loss perturbation and limited attenuation at frequencies of transmission zeros. The resistance per foot of a round conductor at low frequencies is given by [1] as

$$R_{dc} = \frac{10.37 \times 10^{-6} \rho_r}{d^2} \tag{10}$$

where d is the diameter of the conductor in inches and ρ_r is resistivity relative to copper. The resistivity of common conductors relative to solid annealed copper is given in Table 3-1.

At higher frequencies, current flows near the surface of conductors. This phenomenon is known as the skin effect. Consider a round conducting wire. The flux generated by current flowing in the wire is in the form of concentric circles centered in the wire. The inner flux circles link current in the center of the wire but do not link current toward the wire surface, therefore the inductance formed by flux linkage is greater in the center of the wire. This impedes current flow in the wire center and encourages flow toward the surface as the frequency increases.

The skin depth is the conductor penetration depth where the current density has fallen to $1/e$ of the surface value. Most of the current flows within a few skin depths of the surface. The skin depth is

Table 3-1 *Resistivity of conductors relative to copper.*

CONDUCTOR	RELATIVE RESISTIVITY
Silver	0.95
Copper(annealed)	1.00
Copper(hard drawn)	1.03
Gold	1.42
Chromium	1.51
Aluminum	1.64
Beryllium	2.65
Magnesium	2.67
Sodium	2.75
Tungsten	3.25
Zinc	3.40
Brass(66Cu,34Zn)	3.90
Cadmium	4.40
Nickel	5.05
Phosphor Bronze	5.45
Cobalt	5.60
Iron(pure)	5.63
Solder(60/40)	5.86
Platinum	6.16
Tin	6.70
Steel	7.5 to 44
Lead	12.8
Nickel Silver	16.0
German Silver	16.9
Titanium	27.7
Monel	27.8
Constantan(55Cu, 45Ni)	28.5
Kovar A	28.4
Stainless Steel	52
Mercury	55.6
Nichrome	58
Graphite	576

$$\delta = \left(\frac{\lambda}{\pi \sigma \mu c}\right)^{0.5} \tag{11}$$

With μ equal to the free space value, the skin depth in inches is

$$\delta(in) = 2.6\left(\frac{\rho_r}{f}\right)^{0.5} \tag{12}$$

with f in hertz. At higher frequencies, the skin depth becomes quite small and little of the conductor is utilized resulting in an increase is loss resistance. At 1 GHz the skin depth for copper is 0.082 mils.

The high-frequency resistance, R_{ac}, increases with frequency because the current flows in progressively less of the conductor. For a solid, straight, isolated, circular conductor, the high-frequency resistance is

$$R_{ac} = k_f R_{dc} \tag{13}$$

At frequencies where the conductor diameter is less than the skin depth, k_f is nearly one and the dc and ac resistances are equal. When the diameter exceeds about five skin depths

$$k_f = \frac{d}{4\delta} \tag{14}$$

and, therefore, the ac resistance per foot is

$$R_{ac} = 1 \times 10^{-6} \frac{\sqrt{\rho_r f}}{d} \tag{15}$$

For a conductor in proximity to another conductor, or wound in proximity to itself, the mathematics for estimating R_{ac} and inductor Q_u become involved.

Another factor affecting conductor loss is surface roughness. As the frequency is increased, the current flows closer to the surface and the conductor surface roughness eventually exceeds the skin depth. It is not surprising that the resistivity is affected. Conductor loss considering roughness is given by [2]

$$\frac{\rho_{eff}}{\rho} = 1 + \frac{2}{\pi} \tan^{-1}(1.4 \frac{\Delta_{rms}}{\delta}) \qquad (16)$$

where Δ_{rms} is a root mean square measure of the surface roughness assuming roughness ridges are transverse to the current flow and periodic. These assumptions are generally pessimistic and the loss predicted by the previous equation is somewhat greater than measurements reported by several experimenters. Typical roughness figures range from 0.002 mils for metalization on polished ceramic substrates to 0.06 mils for soft substrates. More specific roughness data is supplied by substrate manufacturers.

A plot of conductor loss versus surface roughness is given in Figure 3-1. Notice a rapid transition in loss in the vicinity of roughness equal to the skin depth. For a surface which is smooth in relation to the skin depth, loss is independent of roughness. As the surface roughness substantially exceeds the skin depth, the loss resistance again becomes independent of roughness. Surface treatment is an important factor in loss considerations of microwave filter structures.

3.4 Inductor Technologies

A number of different construction processes are used to create positive reactance elements. Any conductor arrangement which links flux produces inductance, including straight and coiled wire. Coiling the conductor increases flux linkage and therefore inductance. The presence of certain core materials can increase flux concentration and linkage. Short lengths of transmission line are also used to form inductive elements. Next we consider inductive component technologies and their circuit parameters.

3.5 Wire

A straight isolated 22 AWG wire has an inductance of approximately 20 nH per inch of length. The inductance increases with smaller diameter wire. Grover [3] gives formulas for the inductance of various isolated conductors.

Conductors are generally placed in the vicinity of ground or other conductors, which lowers the effective inductance. For example, the 22 AWG wire has an inductance of approximately 26 nH when placed 1 inch above a ground plane, 19 nH at 0.2 inch spacing, and 12 nH when placed on top of 0.062 inch thick

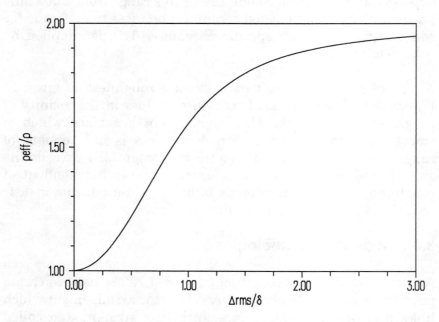

Figure 3-1 *Effective resistivity versus skin-depth and surface roughness.*

PWB with a ground plane. The effective inductance is more accurately predicted using transmission line concepts discussed later in this chapter. Appropriate transmission line models include a wire over ground, wire in a trough, wire between ground planes and others.

The 20 nH per inch approximation is suitable for estimating the additional inductance of leads on inductors or the parasitic lead inductance of capacitors and other components.

3.6 Circular Ring

The inductance of a circular ring (loop) of wire [4] is

$$L(\mu H) = \frac{a}{100}(7.353 \ \log\frac{16a}{d} - 6.386) \tag{17}$$

where a is the mean radius of the ring and d is the diameter of the wire in inches.

3.7 Air Solenoid

An important inductor class is a single layer of wire wound in the form of a cylindrical solenoid with an air core. A rigorous mathematical solution for the inductance of even this simple structure is involved. The inductance is modeled to about 1% by the popular Wheeler [5] formula if the length-to-diameter ratio exceeds 0.33.

$$L(\mu H) = \frac{n^2 r^2}{9r + 10l} \tag{18}$$

where n is the number of turns, r is the radius in inches to the wire center and l is the solenoid length (wire center to wire center) in inches. This model of the solenoid is ideal in that Q_u and parasitics are not considered. A practical inductor model is given in Figure 3-2 where L is given by the above expression.

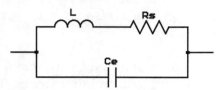

Figure 3-2　*Solenoid inductor model which includes loss resistance and effective parallel self-capacitance.*

The series resistor, R_s, and the solenoid Q_u are related by equation (6).

From equation (2) we see that the reactance and therefore Q_u should increase linearly with frequency and from equation (15) we see that conductor ac resistance increases with the square root of frequency. Although equation (15) is valid only for an isolated conductor, we may hypothesize a similar relationship as a function of frequency for the solenoid. We would therefore expect Q_u to increase as the square root of frequency. As it turns out, the unloaded Q of a single-layer copper-wire air solenoid has been empirically found [6] to be

$$Q_u \approx 190d\psi\sqrt{f} \tag{19}$$

where d is the solenoid diameter in inches, f is frequency in megahertz and ψ is a function of l/d and the wire diameter to wire spacing ratio, d_w/s. The d_w/s resulting in the maximum Q_u ranges from about 0.6 to 0.9 for normal l/d and is plotted in Figure 3-3. Given a near optimum wire spacing, a curve fit which fits Medhurst's data for ψ_{opt} to a few percent is

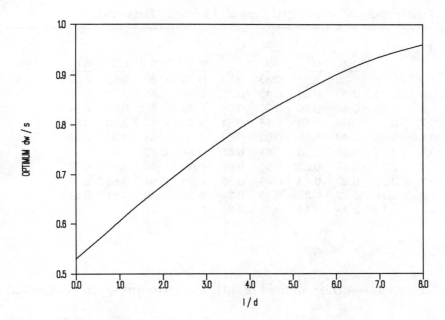

Figure 3-3 *Optimum wire spacing to wire diameter ratio versus solenoid length to diameter ratio.*

$$\psi_{opt} = 0.9121 - \frac{0.4964}{\sqrt{l/d}} + \frac{0.0709}{l/d} \tag{20}$$

With $l/d = 2$, ψ_{opt} is approximately 0.79. In the above expression for Q_u, ψ is constant for a given l/d. We will see later there is an optimum l/d value for maximum Q_u. Since ψ is a constant and the frequency is a required value, the only parameter under the designer's control is the coil diameter. Increased solenoid diameter increases Q_u.

ψ for a range of d_w/s and l/d is given in Table 3-2. Notice that as the solenoid length is increased, d_w/s for maximum ψ increases as shown in Figure 3-3. Also, the table suggests that the highest Q_u is achieved with long solenoids since ψ increases with increasing l/d for all winding spacings. However, another

62 HF Filter Design and Computer Simulation

Table 3-2 ψ versus l/d and d_w/s from Medhurst.

d_w/s					l/d				
	0.2	0.4	0.6	0.8	1.0	2.0	4.0	6.0	8.0
1.0	0.18	0.26	0.31	0.35	0.38	0.63	0.80	0.89	0.93
0.9	0.24	0.33	0.39	0.43	0.47	0.69	0.84	0.90	0.93
0.8	0.28	0.40	0.46	0.50	0.55	0.75	0.87	0.90	0.91
0.7	0.32	0.46	0.53	0.58	0.61	0.78	0.87	0.90	0.89
0.6	0.34	0.49	0.57	0.63	0.67	0.78	0.85	0.87	0.86
0.5	0.34	0.48	0.56	0.61	0.65	0.74	0.80	0.81	0.82
0.4	0.31	0.45	0.52	0.56	0.60	0.69	0.74	0.75	0.76
0.3	0.25	0.37	0.44	0.49	0.52	0.50	0.64	0.66	0.67
0.2	0.19	0.27	0.33	0.36	0.39	0.45	0.49	0.51	0.51
0.1	0.10	0.14	0.17	0.19	0.21	0.25	0.27	0.28	0.29

factor causes the solenoid length for optimum Q_u to be shorter.

The solenoid winding consists of conductors separated from each other by an air dielectric, thus forming capacitance. The winding also has capacitance to the ground plane. This capacitance is effectively modeled as capacitance, C_e, in parallel with the inductance and conductor loss. Medhurst also empirically determined the value of this effective capacitance for an unshielded single-layer solenoid with one end grounded. He found

$$C_e \approx 2.54Hd \qquad (21)$$

where C_e is in picofarads, H is a function of the length to radius ratio and d is again in inches. Table 3-3 gives values of H versus l/d. Notice that the effective capacitance is only a function of the diameter and l/d. Intuition suggests that as d_w/s increases and the windings approach each other, the interwinding capacitance would increase. As the spacing decreases to zero, the capacitance would approach infinity. However, as the windings approach each other, the flux from the current in adjacent windings forces the current in the wires away from each other and toward the inside of the solenoid. The effective spacing between the current filaments becomes the wire center-to-center spacing and not the closing edge-to-edge spacing. This factor also explains why the ac resistance of a coiled wire is higher and more complex to determine mathematically than for an isolated conductor.

From Table 3-3 it is evident, that for a length to diameter ratio from 0.5 to 2.0, a value of 0.48 for H is within $\pm$ 2% and

$$C_e \approx 1.22d \tag{22}$$

At low frequencies, C_e has little effect. As the operating frequency increases, C_e increases the effective inductance and decreases the unloaded Q. The resonant frequency of the inductor model in Figure 3-2 is

$$f_r = \frac{1}{2\pi\sqrt{LC_e}} \tag{23}$$

As the resonant frequency is approached, the reactance increases which effectively increases the inductance. Above resonance, the reactance of the solenoid becomes capacitive and begins decreasing.

The parasitic capacitance also reduces the Q_u. The effective Q_u as reduced by the parasitic capacitance is approximately

$$Q_u \approx Q_{uo}\left[1 - \left(\frac{f}{f_r}\right)^2\right] \tag{24}$$

where Q_{uo} is the unloaded Q without parasitic capacitance and f is the operating frequency. As f approaches f_r, Q_u approaches zero.

Table 3-3 *Solenoid C_e factor, H, versus the l/d due to Medhurst.*

l/d	H		l/d	H
0.10	0.96		1.50	0.47
0.15	0.79		2.00	0.50
0.20	0.70		3.00	0.61
0.30	0.60		4.00	0.72
0.40	0.54		5.0	0.81
0.50	0.50		7.0	1.01
0.70	0.47		10.0	1.32
0.90	0.46		15.0	1.86
1.00	0.46		20.0	2.36

Equation (19) suggests that an inductor with arbitrarily large Q_u exists for a suitably large diameter. However, increasing the radius increases the parasitic capacitance and restricts the size, therefore the maximum Q_u.

Plotted in Figure 3-4 are the effective inductance, L_e, and effective unloaded Q, Q_e, as a function of frequency for a copper-wire solenoid inductor with a length to diameter ratio of three. The inductance is constant and the Q increases with frequency until the effects of the effective parallel capacitance become significant. Curves (a) are with a value of effective parallel capacitance estimated by Medhurst and curves (b) are with a safety factor of four times Medhurst's effective capacitance. At the self-resonant frequency, the effective Q is zero. Notice that Q_e is maximum at an operating frequency well below the self-resonant frequency. Also notice that the effective inductance increases as the self-resonant frequency is approached and becomes infinite at the self-resonant frequency.

Medhurst's estimate of C_e is based on an inductor whose axis is perpendicular to a ground plane with one end grounded. A nearby shield or ground-plane can significantly increase the effective parallel capacitance and lower the self-resonant frequency. Notice the effect of attempting to design an inductor

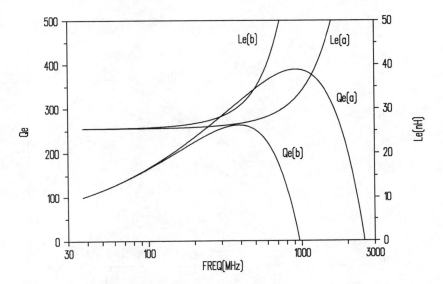

Figure 3-4 *Solenoid L_e and Q_e including parasitic capacitance, C_e. (a) is with C_e by Medhurst and (b) is with a safety factor of 4 for C_e.*

for maximum Q at the desired operating frequency if the capacitance is greater than expected. A small inductor diameter avoids self-resonance but reduces the available Q.

Given in Figure 3-5 is the effective unloaded Q, Q_e, versus frequency for solenoid length to diameter ratios of 1, 2, 3 and 10. The number of turns is adjusted to maintain an inductance of approximately 25.5 nH. The wire gauge is adjusted to maintain a near optimum wire diameter to wire spacing ratio as per Medhurst. At low frequencies, Q_e increases with longer solenoids. However, longer solenoids have increased self-capacitance and a lower self-resonant frequency. From these curves it is clear that the optimum length to diameter ratio is rather broad, perhaps being centered on approximately 3:1. If size constraints limit the solenoid size to less than the maximum value, longer solenoids are indicated.

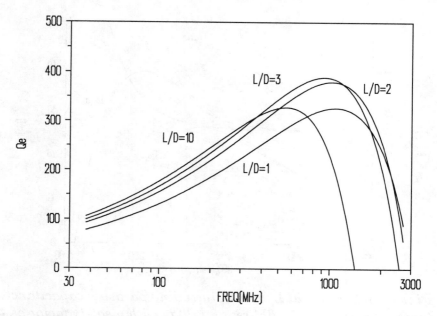

Figure 3-5 Q_e *for l/d ratios of 1, 3 and 10. The number of turns is adjusted to maintain 23.5 nH. The wire gauge is adjusted to maintain a Medhurst optimum diameter to spacing ratio.*

Given in Figure 3-6 is the effective unloaded Q, Q_e, versus frequency for copper-solenoid inductors with diameters of 0.2, 0.1 and 0.05 inches. The length to diameter ratio is 3 to 1 and the wire gauge is 22. The turns are adjusted to fill the winding length with the optimum wire diameter to wire spacing of approximately 0.75 for a length to diameter ratio of 3 to 1, resulting in 18, 9 and 4.5 turns for the 0.2, 0.1 and 0.05 inch diameter solenoids, respectively.

At the lower frequencies, Q_e increases in direct proportion to the wire diameter. At higher frequencies, the larger diameters are unsuitable. The maximum available Q is relatively constant with frequency, providing smaller diameters are used.

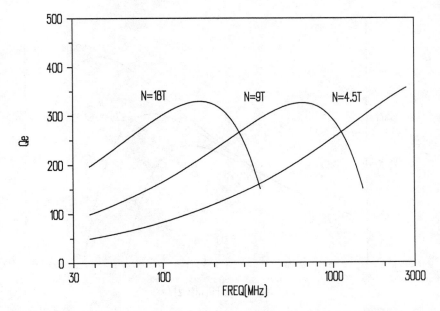

Figure 3-6 *Effective unloaded Q for solenoids with diameters of 0.2 (18T), 0.1 (9T) and 0.05 (4.5T) inches. The number of turns are adjusted for optimum spacing with 22 gauge wire.*

Given in Figure 3-7 is the effective Q of copper-wire solenoids versus wire gauge with a length to diameter ratio of 3 and a diameter of 0.1 inches. The number of turns is adjusted to maintain the optimum spacing. The approximate inductances are 142, 59 and 20 nH and 14, 9 and 5.2 turns respectively for the 26, 22 and 18 gauge cases.

At low frequencies, the wire gauge, and therefore the number of turns, may be selected to realize the desired inductance. At high frequencies, the higher inductance ssociated with an increased number of turns decreases the resonant frequency and degrades the unloaded Q. Therefore, for a given allowable diameter, to obtain the maximum unloaded Q as the frequency is increased the wire diameter is increased while maintaining the optimum spacing. This reduces the number of turns and the inductance.

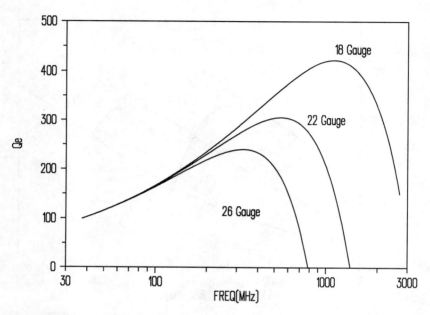

Figure 3-7 *Effective unloaded Q versus frequency for solenoids with l/d=3 and a diameter of 0.1 inches. The number of turns (#26=14T, #22=9T, #18=5.2T) are adjusted for optimum spacing.*

Given in Figure 3-8 is the effective unloaded Q of copper-wire solenoids with a length to diameter ratio of three and a diameter of 0.1 inches. The wire gauge is 22 and the spacing is adjusted to achieve 10.8, 9 and 4.5 turns and approximately 85, 59 and 15 nH inductance, respectively. At lower frequencies, the optimum number of turns for unloaded Q is 9 which corresponds to Medhurst's optimum spacing. At higher frequencies, a fewer number of turns decreases the inductance and raises the self-resonant frequency resulting in increased unloaded Q.

3.8   Solenoid With Shield

Solenoids produce magnetic flux which has the potential to link with adjacent solenoids or conductors and therefore provide undesired transmission paths which affect the filter response. Electrostatic coupling also occurs between solenoids when the

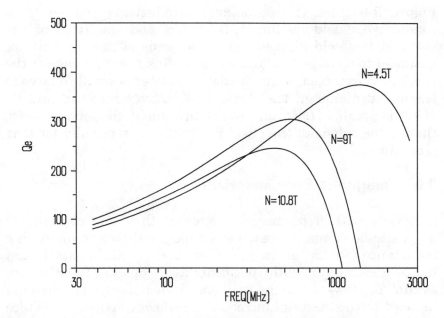

Figure 3-8 *Effective unloaded Q of 22 gauge copper-wire solenoids with l/d=3. The number of turns is varied and the wire is spread as necessary to occupy the 0.3 inch length.*

voltage potential on one solenoid induces a voltage potential on a second solenoid via mutual capacitance. To avoid these difficulties, solenoids are enclosed within a housing of good conductivity or high permeability. At low frequencies, shields with a high permeability are used. At high frequencies, the eddy currents induced in conducting shields effectively block both magnetic and electrostatic fields.

Shielding a solenoid increases C_e, lowering the resonant frequency, and increasing the loss resistance, thus lowering Q_u. Higher shield conductivities reduce shield losses and minimize Q_u degradation. The solenoid inductance is decreased when non-magnetic shield material is used and is increased when magnetic materials are used. Greater shield spacings from the solenoid minimize all of the above effects.

Figure 3-9 gives the reduction in inductance caused by a conductive shield as functions of l/d and the ratio of the solenoid to shield diameter [7]. The ends of the solenoid are assumed to be at least one solenoid radius from the ends of the shield. As expected, if the solenoid diameter is small relative to the shield diameter, the shield is effectively removed and its effect is small. However, even if the shield diameter is twice that of the solenoid, the effect is significant especially for long l/d ratios.

3.9   Magnetic-Core Materials

Materials with a permeability greater than one concentrate magnetic fields and increase flux linkage, therefore increasing inductance for a given inductor size. Silicon-steel and nickel-iron alloys with permeabilities up to 100,000 are available. However, as the frequency is increased, eddy currents induced in the material introduce significant losses. To reduce eddy-current losses, the core is subdivided by winding tape as thin as a thousandth of an inch or less. Even so, above about 1

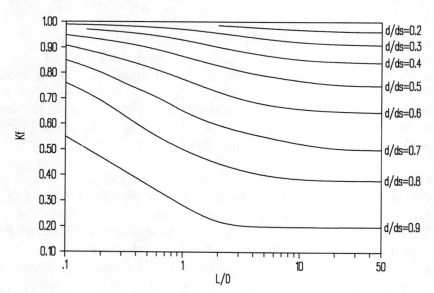

Figure 3-9 K_f versus solenoid length to diameter ratio.

MHz, eddy-current losses become prohibitive. At higher frequencies, a more successful strategy for preventing eddy-currents is to form cores by pressing together magnetic-material powders in an insulating binder. Representative commercial core materials for high-frequency use, their permeability relative to air, μ_r, and useful operating frequency range are given in Table 3-4.

While magnetic core materials allow greater inductance with less wire and smaller physical size, other limitations are introduced. The temperature stability of the material permeability affects the temperature stability of the inductance. In air, the flux produced, B, is linearly proportional to the magnetizing force, H. The magnetizing force is the ampere-turns divided by the magnetic path length. Most magnetic materials possess hysteresis, where the produced flux is a function of both the present and past magnetizing force. These non-linear effects introduce signal harmonics and intermodulation distortion. Also, there is a maximum value of induced flux beyond which an increase in magnetizing force causes little increase in flux, limiting the maximum useful application power level.

3.10 Solenoid with Core

Cylindrical cores longer than a solenoid winding are used to increase inductance for a given physical size. Cylindrical cores shorter than the solenoid winding and moved along the winding axis are used for inductance tuning with the greatest inductance occurring with the core centered.

The effective permeability of a core, μ_e, is the factor by which the inductance is increased over the value with an air core. For a cylindrical core, the effective permeability is as much a function of the l/d ratio of the core as it is the initial relative permeability of the core, μ_r. The effective permeability is related to the material permeability by the relation [8]

Table 3-4 *Representative list of commercial powdered-iron and ferrite high-frequency magnetic core materials for relative permeability from 1 (phenolic) through 2500.*

μ_r	Amidon	Fair-Rite	Ferrox-cube	Indiana General	Mag-netics	Micro-metals	Stack-pole
1						#0	
4						#12, #17	
6						#10	
7.5							C/14A
8.5						#6	
9						#4, #7	
10						#2	
12.5							C/14
16				Q3			
20		#68				#1	
25						#15	
35						#3, #8	C/12
40	FT-63	#63, #67		Q2		#42	
75						#41	
100		#65					
125	FT-61	#61	4C4	Q1			C/11
175		#62					
250	FT-64	#64					
300		#83					
375		#31					
400				G			
500							C/5N
750			3D3		A		
800		#33					
850	FT-43	#43		H			C/7D
950				TC-3			
1200		#34					
1400					C		
1500				TC-7			
1800			3B9				
2000	FT-77	#77			TC-9	S, V, D	
2200				05			
2300			3B7		G		
2500	FT-73	#73		TC-12			

$$\mu_e = \frac{\mu_r}{1 + N_f(\mu_r - 1)} \tag{25}$$

where N_f is the demagnetization factor. A curve fit to the demagnetization factor data in Snelling is

$$N_f = \frac{0.28}{l/d} - 0.0158 - \frac{0.0915}{\mu_r} + \frac{0.00063\, l/d}{\log(\mu_r)} \tag{26}$$

where l/d is the length to diameter ratio of the core, not the winding. The validity ranges are $1 < l/d < 100$, $1 < \mu_r < 1000$ and a winding which is evenly spaced and occupies 85 to 95% of the core length. For example, μ_e of a cylindrical core with μ_r equal to 125 and an l/d ratio of 4 is 16.1 and with an l/d ratio of 8 is 34.9. Plots of μ_e versus μ_r for various l/d is given in Figure 3-10 for solenoids which are evenly spaced and occupy 85 to 95% of the length of the core. Notice the effective permeability is almost independent of the material permeability for short l/d. Because the effective permeability is nearly independent of the material permeability, the temperature instability of the core material is reduced.

A metallic conductive core, such as aluminum, reduces inductance and is used to tune inductor solenoid values as much as 40%. The Q is reduced typically from 25 to 50%, which at very high frequencies may be less degradation than produced by magnetic material cores.

3.11 Toroid

Toroid shaped cores realize much of the relative permeability of the core material. They also tend to confine the magnetic fields within the toroid and therefore minimize shielding requirements. The effective permeability is somewhat below the material permeability and inductance is tuned slightly by compressing or

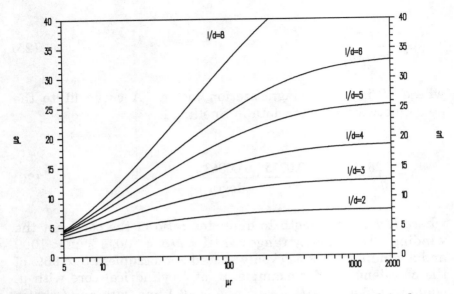

Figure 3-10 *The effective permeability of a cylindrical core versus the material permeability and the core length to diameter ratio.*

spreading the winding evenly about the core. Compressed windings yield higher inductance.

The inductance of toroid windings is approximately

$$L(nH) \approx 11.7\mu_r t N^2 \log\left(\frac{OD}{ID}\right) \tag{27}$$

where t is the core thickness in inches, N is the number of turns (passes through the center of the core), OD is the core outer diameter and ID is the core inner diameter. The turns are assumed evenly spaced around the core and the wire diameter is small and tightly conformed to the core.

The initial relative permeability of representative commercial high-frequency core materials is given in Table 3-4. Commercial toroid core dimensions in inches are given in Table 3-5.

Table 3-5 *Representative list of commercial toroid core dimensions.*

OD	ID	Thick	Amidon	Fair-Rite	Ferrox cube	Indiana General	Mag-netics	Micro metal
0.100	0.050	0.050					40200	
0.100	0.070	0.030		701		F426-1		
0.125	0.062	0.050						
0.160	0.078	0.060						T-12
0.190	0.090	0.050			213T050			T-16
0.200	0.088	0.070						
0.230	0.120	0.060	FT-23	101	1041T060	F303-1	40601	T-20
0.230	0.120	0.120		901				
0.255	0.120	0.096						
0.300	0.125	0.188				F867-1		T-25
0.307	0.151	0.128						
0.375	0.187	0.125	FT-37	210	266T125	F625-9	41003	T-30
0.375	0.205	0.128						
0.440	0.229	0.159						T-37
0.500	0.281	0.188	FT-50	301	768T188			T-44
0.500	0.303	0.190						
0.500	0.312	0.250		1101		F627-8	41306	T-50
0.500	0.312	0.500		1901				
0.690	0.370	0.190						T-68
0.795	0.495	0.250						T-80
0.825	0.520	0.250	FT-82	601				
0.825	0.520	0.468		501				
0.870	0.500	0.250		401				
0.870	0.540	0.250		1801	846T250	F624-19	42206	
0.942	0.560	0.312						T-94
1.000	0.500	0.250		1501		F2070-1	42507	
1.000	0.610	0.250		1301				
1.060	0.570	0.437						T-106
1.142	0.748	0.295	FT-114	1001	K300502		42908	
1.225	0.750	0.312		1601				
1.250	0.750	0.375		1701		F626-12		
1.300	0.780	0.437						T-130
1.500	0.750	0.500			528T500		43813	
2.000	1.250	0.550						T-200
2.000	1.250	0.750			400T750			

The unloaded Q of toroid inductors is as difficult to compute as it is for the simple solenoid, and when a magnetic core material is used, core losses must be considered as well. Core losses are represented as an additional parallel resistance in the equivalent circuit model.

Empirical data for Q_u versus frequency for representative toroids for high-frequency applications are given in Figure 3-11, courtesy of Micrometals [9], which includes many other curves. The curves selected here are among the highest Q for the represented frequency range. Curves a through d are toroids with magnetic core material. Notice the downward trend in maximum Q versus frequency. It appears the Q would vanish at a frequency of 300 MHz. Indeed, due to core losses, magnetic materials are seldom used above 200 to 300 MHz.

Curves e and f are toroid windings on non-magnetic (phenolic) material, Micrometals material #0. If toroid e was "straightened" into a solenoid, the "square diameter" would be 0.25×0.44 inches and the "length" would be 2.56 inches. The inductance would be approximately 26.8 nH, one-fifth the inductance of the toroid form. Unfortunately the self capacitance of toroids is greater than solenoids, because the ends, which are at the maximum potential difference, approach each other as the windings fill the toroid. Because self-capacitance is less significant at lower frequencies, and because lower frequencies require larger inductance for a given reactance, the increased inductance of toroid windings is most beneficial at lower frequencies. Optimum toroid design shares an important fundamental concept with solenoid design; for best Q the maximum physical size consistent with the peak Q and inductance tolerance is used. With increasing frequency, that maximum physical size must decrease.

3.12 Capacitors

The basic capacitor consists of two conducting plates separated by a dielectric material. The capacitance in farads is given by

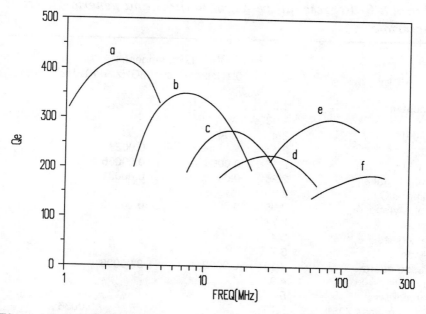

Figure 3-11 *Q versus frequency of toroid inductors on various Micrometals cores. Curves a through d are powdered iron. Curves e and f are phenolic. See text for details.*

$$C = \frac{8.85 \times 10^{-12} \epsilon_r A}{S} \tag{28}$$

where ϵ_r is the relative dielectric constant, A is the plate area in square meters and S is the plate spacing in meters. With dimensions in inches

$$C(pF) = \frac{.2248\ \epsilon_r A}{S} \tag{29}$$

The relative dielectric constant of air is near one, 1.0006. The relative dielectric constant of representative natural and generic

Table 3-6 *Representative list of natural and generic dielectric materials.*

MATERIAL	ε_r	Loss tangent			
		0.1GHz	1GHz	3GHz	10GHz
Vacuum	1.0				
Air	1.0006				
Wood	2-4	0.04		0.03	
PTFE	2.1			0.0028	
Vaseline	2.16	0.0004		0.00066	
Polyethylene	2.26	0.0002		0.00031	
Mineral Oil	2.4				
Polystyrene	2.55	0.0001		0.00033	
Paper	2.7+				
Plexiglass	3.4				
Nylon	3.7				
Fused quartz	3.78	0.0002		0.00006	
Bakelite	4.9				
Pyrex glass	5				
Corning glass 7059	5.75				0.0036
Porcelain	6				
Beryllium oxide	6.3				0.006
Neoprene	6.7				
Silicon	11.7		0.005		0.015
Gallium Arsenide	13.10				0.0016
Water, distilled	78	0.005		0.157	
Generic PWB materials, approximate values					
G-10/FR4, low resin	4.9	0.008			
G-10/FR4, high resin	4.2	0.008			
Paper/Phenolic	4.8	0.03			
Alumina, 96%	10		0.0002		
Alumina, 99.5%	9.6	0.0002			0.0003

materials is given in Table 3-6. Any insulator can be used as a dielectric. The ideal dielectric has low loss for maximum unloaded capacitor Q, a large relative dielectric constant for small physical capacitor size, excellent temperature stability, and a high-breakdown voltage. The first high-quality dielectrics were natural materials such as mica and quartz. Titanate materials with very high dielectric constants have been used for some time, but the temperature stability and loss increase with increasing dielectric constant. PTFE and other plastics with good stability and quality have been used for decades at microwave frequencies. More recently, ceramic materials have been developed which possess excellent stability and quality and also have higher dielectric constants. The relative dielectric constant of representative commercial materials in Table 3-7.

Losses occur in both the conductors and dielectric of capacitors and printed circuit boards. These losses limit the available unloaded Q. That component of the unloaded Q attributed to dielectric loss is Q_d. Q_d is related to the loss tangent of the dielectric material by the simple relation

$$Q_d = \frac{1}{\tan\delta} \tag{30}$$

A simple but practical capacitor model is given in Figure 3-12. The dielectric loss is represented by R_p. At high frequencies and high-quality dielectrics, R_p is generally insignificant and the majority of the loss is conductor losses modeled by R_s.

An important parasitic of capacitors is the series inductance produced by flux linkage in the leads and the spreading currents of the plates. Even chip capacitors have series inductance. At higher frequencies, the inductive reactance becomes significant and reduces the total reactance, causing the effective capacitance to increase. At resonance, the impedance is resistive, and above resonance the reactance is inductive and increases with frequency.

A rough estimate of the series inductance is 20 nH per inch of conduction and displacement current path length. For example the inductance of a chip capacitor 0.05 inches long is approximately 1 nH while a monolithic capacitor with 0.2 inch lead spacing has approximately 0.4 inches of effective length and a series inductance of 8 nH. A 20 pF chip capacitor with 1 nH lead inductance has an effective capacitance increase of 10% to

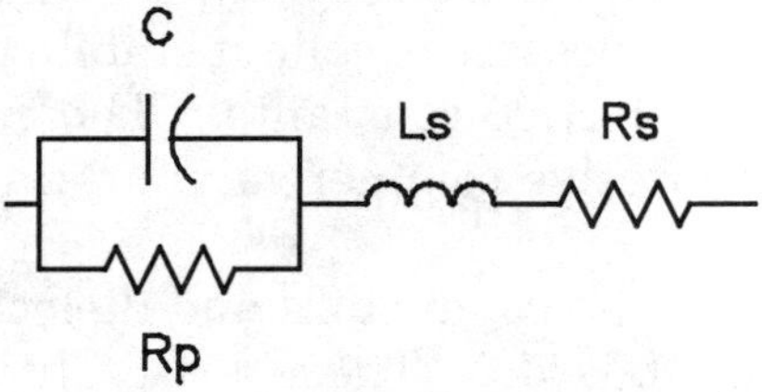

Figure 3-12 Equivalent circuit model for capacitors. At high frequencies with high-quality dielectrics, R_p is insignificant.

22 pF by 127 MHz and a series resonance of 1.13 GHz. Wider conductors, and therefore wider chip capacitors, have reduced inductance while longer capacitors have increased inductance.

At VHF frequencies and lower, the unloaded Q of high frequency capacitors can be several thousand. Inductive element unloaded Q is typically lower so capacitor unloaded Q is often ignored. At UHF and microwave frequencies, more careful capacitor selection is advised. Lower value capacitors typically yield higher unloaded Q and a higher internal filter impedance is used to reduce capacitor values.

The unloaded Qs of 1, 10 and 100 pF AVX AQ11 chip capacitors [10] are plotted in Figure 3-13. From equation (7) we see that the unloaded Q of a capacitor is inversely proportional to frequency because the reactance is decreasing. R_s for capacitors is referred to as the effective series resistance, *ESR*. The *ESR* for high-quality chip capacitors is typically less than 0.1 ohm and decreases with decreasing frequency. With increasing frequency, the series inductance of the capacitor introduces positive reactance which cancels some of the capacitive reactance, further decreasing the Q. All of these factors

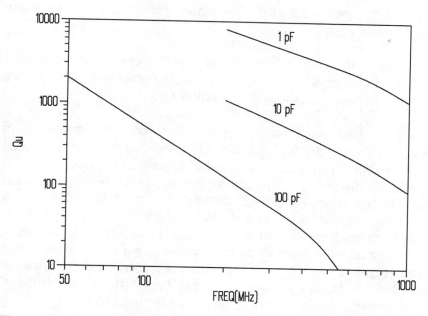

Figure 3-13 *Unloaded Q of AVX AQ11 1 pF, 10 pF and 100 pF chip capacitors versus frequency.*

contribute to a rapid decrease in capacitor Q with increasing frequency. The downward turn in Q versus frequency near the high-frequency end of the curves is due to the effects of the series inductance. All of these effects are observed in other capacitor types such as leaded capacitors and trimmer capacitors. At higher frequencies, filter structures which require small values of capacitance are imperative if high unloaded capacitor Q is to be achieved.

3.13 Transmission Lines

Inductors and capacitors are pure elements in the sense that energy is stored in either magnetic or electrostatic fields, not both. This offers the advantage of wide bandwidth, the frequency range over which the element properties are retained. The bandwidth is restricted only by parasitic considerations.

Careful design and material selection assist in reducing these parasitic effects.

To limit inductor self-capacitance, as the operating frequency increases the physical size of inductors is decreased which places an upper limit on available unloaded Q.

Transmission lines store energy in both magnetic and electrostatic fields. An equivalent lumped-element circuit model includes both inductance and capacitance, and resistors when losses are included. The transmission line is continuous, and may be represented as a distributed circuit model with an infinite number of series inductors and shunt capacitors. Inductors and capacitors are referred to as lumped elements, and transmission lines as distributed elements. A transmission line segment which is electrically shorter than a quarter wavelength may be modeled accurately with three lumped elements and less accurately with one lumped element.

Increasing the physical size of distributed elements provides for increased energy storage without a commensurate increase in losses. Therefore, by making distributed elements larger than corresponding lumped elements, increased unloaded Q is available. The maximum physical size and Q is limited only by moding.

3.14 Modes

Two-conductor transmission line structures in a homogeneous dielectric media, such as coaxial cable and stripline, are normally operated in the transverse electromagnetic (*TEM*) propagation mode. The electric and magnetic fields are perpendicular to each other and to the direction of the conductors. In such media, the propagation velocity is equal to the velocity of light in the dielectric material and is proportional to the square root of the relative dielectric constant. It is frequency independent.

When the cross section of the transmission line is appreciable with respect to a wavelength, other propagation modes are possible, including transverse electric (TE) or transverse magnetic (TM). TE and TM are the desired propagation mode in transmission lines such as waveguide. These modes are dispersive; they are frequency dependent. While such modes are the basis of waveguide, the existence of these propagation modes in coax is undesirable. The first higher-order mode in coax is a TE mode which is supportable when the average circumference of the line is approximately equal to the wavelength in the coax dielectric material. The cutoff frequency is approximately

$$f_c(MHz) \approx \frac{7510}{\sqrt{\epsilon_r}} \frac{1}{(a+b)} \tag{31}$$

where a and b are the inner and outer diameter in inches of the dielectric.

In two-conductor transmission lines with non-homogeneous dielectric media, such as microstrip and coplanar waveguide, the propagation mode is quasi-TEM. A portion of the electric fields are in the dielectric between the strip and the ground plane while other fields exist in the region above the strip with air as a dielectric. At frequencies where the electrical distance in the dielectric material between the strip and ground plane is much less than a wavelength, microstrip behaves as a non-dispersive TEM line. With increasing frequency, as the substrate thickness of microstrip becomes appreciable, the propagation velocity and the characteristic impedance of the line increase.

Another factor which limits the practical maximum physical size of transmission line structures is the presence of discontinuity effects, which are discussed in more detail later in this chapter.

3.15 Transmission Line Unloaded Q

For a given operating frequency, the physical size of a distributed element is limited by higher-mode cutoff and discontinuity effects, and can be larger than lumped inductors whose size is limited by parasitic capacitance. Therefore, the achievable unloaded Q of distributed elements is higher than with lumped elements, providing size is not a constraint. The unloaded Q of distributed resonators, equal to a multiple of a quarter-wavelength long, or which is very long, is

$$Q_u = \frac{\pi \lambda_{go}}{\alpha_t \lambda_o^2} \tag{32}$$

where λ_{go} is the wavelength in the line, λ_o is the wavelength in air and α_t is the line attenuation in nepers per unit length. The unloaded Q increases with increasing size in transmission line elements because the attenuation decreases with increasing size.

Losses in transmission lines occur in both the dielectric and the conductors. When the transmission line length is other than a multiple of a quarter-wavelength, equation (32) fails to represent the balance of these losses. For example, lines which are less than one quarter wavelength have lower unloaded Q than is predicted by the equation if α_t is not appropriately adjusted. It has been reported (see Section 8-14) that the unloaded Q of a transmission-line element shorter than a quarter wavelength is

$$Q_u = Q_u \sin^2\theta \tag{33}$$

where θ is the electrical line length. Intuitively this expressive would seem to assume all losses are conductor losses. If dielectric losses are significant, this expression may be invalid.

3.16 Coupled Transmission Lines

When transmission line conductors share a common ground reference and are placed in close proximity so their fields interact, signals in each line may be induced in the other. This may be desired, as in couplers, or undesired, creating interference or crosstalk. The behavior of two coupled lines is characterized by not one but two characteristic impedances. The even-mode impedance, Z_{oe}, is defined by exciting both conductors with the same signal amplitude and phase. The odd-mode impedance, Z_{oo}, is defined by exciting the conductors with signals of equal amplitude, but 180 degree out of phase.

The maximum coupling between lines occurs when the coupled line section is 90 degrees long. The coupling coefficient, k, is related to the even and odd-mode characteristic impedance by

$$k = \frac{Z_{oe} - Z_{oo}}{Z_{oe} + Z_{oo}} \tag{34}$$

The coupler is matched at its ports to the system characteristic impedance when

$$Z_o = \sqrt{Z_{oe} Z_{oo}} \tag{35}$$

From these expressions we may also derive

$$Z_{oe} = Z_o \sqrt{\frac{1+k}{1-k}} \tag{36}$$

$$Z_{oo} = Z_o \sqrt{\frac{1-k}{1+k}} \tag{37}$$

At full coupling, the coupling in decibels is

$$k(dB) = 20 \log k \tag{38}$$

3.17　Transmission-Line Elements

At higher frequencies, transmission lines are useful filter elements. From a lumped element perspective, lines electrically shorter than 90 degrees are used as capacitive or inductive reactors and lines which are a multiple of a quarter wavelength long act like series or parallel L-C resonators.

Filters constructed with distributed elements are designed either by beginning with lumped element filter theory and converting the lumped elements to equivalent distributed elements, or by employing synthesis theories developed specifically for distributed structures. Before we consider these filter design approaches, we will investigate distributed elements in more detail.

Consider a transmission line terminated with a short to ground. This zero impedance termination is plotted on the Smith chart in Figure 2-3 at the far left end of the real axis at the circumference of the chart. The impedance looking into this shorted line graphs on the circumference of the chart in a clockwise direction with increasing electrical length. One complete revolution around the Smith chart is 180 degrees. The impedance looking into a 50 ohm shorted line which is 45 degrees long at a given frequency graphs on the circumference directly above the center of the chart. This point is on a zero resistance circle and a normalized reactance arc of $+j1$, or 50 ohms positive reactance. Shorter length lines have a smaller reactance and longer length lines, less than 90 degrees, have a larger reactance. The impedance looking into a shorted, lossless, transmission line is purely reactive and

$$X = Z_o \tan\theta \tag{39}$$

where Z_o is the transmission line characteristic impedance and θ is the electrical length of the line. For $\theta < 90$ degrees, the reactance is positive and the shorted line looks inductive.

If the electrical length is much shorter than 90 degrees at the highest frequency of interest, the tangent function is nearly linear and the reactance increases linearly with frequency much like an inductor. The reactance of a shorted 80 ohm transmission line which is 22.5 degrees long at 1 GHz is plotted versus frequency in Figure 3-14. The reactance of a 5.27 nH ideal inductor and the same inductor with 0.1 pF self-capacitance are also plotted in Figure 3-14. The inductor is a small 0.03 inch diameter solenoid with C_e as predicted by Medhurst at 0.06 pF, so 0.1 pF represents a small safety factor. Notice that below 1 GHz all three are roughly equivalent.

A transmission line which is 22.5 degrees long at 1 GHz is 90 degrees in length at 4 GHz and from equation (39), the line reactance becomes infinite. Losses in real lines moderate the impedance by introducing resistance. The 5.27 nH inductor with

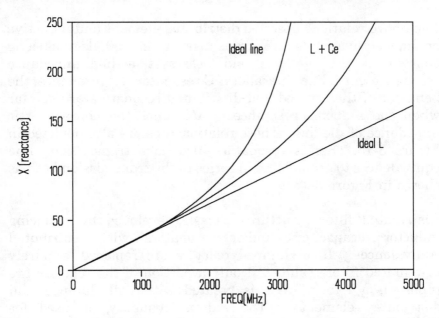

Figure 3-14 *Reactance vs. frequency for an ideal 5.27 nH inductor (Ideal L), this inductor with C_e=0.1 pF (L+Ce) and an ideal 80 ohm shorted line which is 22.5 degrees long at 1 GHz (Ideal line).*

0.1 pF of effective parallel capacitance resonates at 6.93 GHz, significantly higher than the resonate frequency of the line, and therefore the real inductor retains useful inductor behavior over a wider bandwidth.

3.18   Lumped-Distributed Equivalences

We have seen that an inductor and a shorted transmission line are similar. There are other transmission line equivalences. A transmission line open at the opposite end looks much like a capacitor if the electrical length at the highest frequency is less than approximately 22 degrees. A series inductor is approximated by a short length, high characteristic impedance, series transmission line and a shunt capacitor is approximated by a short length, low impedance, series transmission line. These relations are depicted in Figure 3-15.

The above relations describe distributed element and inductive or line equivalences. A transmission line 90 degrees long shorted at the opposite end possesses a high-impedance resonance much like a parallel L-C resonator. At resonance, the behavior of the lumped and distributed resonators are similar when the inductor reactance is $4/\pi$ times the characteristic impedance of the line. These relationships are also depicted in Figure 3-15. Series resonators also have transmission line equivalences to ground and in series to the transmission path as shown in Figure 3-15.

Distributed filter structures can be developed by replacing inductors, capacitors and/or resonators with distributed equivalences. If all lumped reactors are replaced, a purely distributed filter results. If only a portion of the reactors are replaced, a hybrid filter is created with both lumped and distributed elements. The radian frequency, ω, used for equivalent calculations in Figure 3-15 is the cutoff frequency for lowpass and highpass filters and the center frequency for bandpass and bandstop filters.

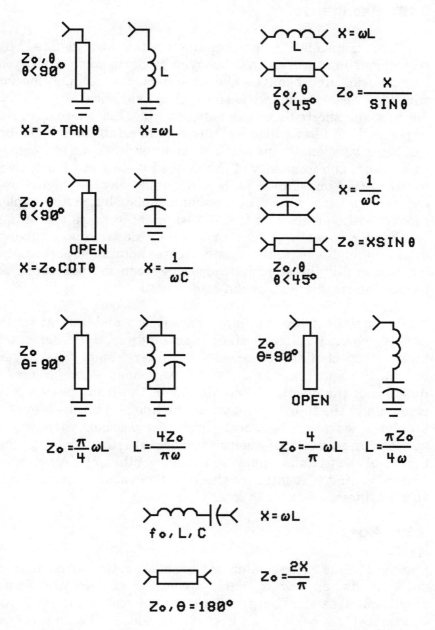

Figure 3-15 *Lumped and distributed (transmission line) element approximate equivalent relationships.*

3.19   Reentrance

Because transmission line characteristics are defined by trigonometric functions, the behavior is harmonic. A shorted transmission line behaves like an inductor at frequencies for which the electrical length is short. At an electrical length of 90 degrees, the shorted line resonates and possesses a high input impedance. A filter which has a passband established by using a quarter wavelength line will have another passband at roughly three times the frequency of the desired passband. For longer lengths, the impedance is periodic and may be high, low, inductive or capacitive. The passband of distributed filters may reoccur at frequencies as low as two or three times the initial passband frequency, totally or severely destroying stopband attenuation. Reentrance not only creates additional passbands, but it also fills the stopband region between the passbands to reduce intermediate stopband attenuation.

While parasitics are an important aspect and limitation in lumped element filter design, reentrance is a fundamental limiting aspect of distributed-element filter design. The degree to which reentrance limits filter performance is largely a function of the electrical line lengths of the transmission line elements in the filter. The shorter the line length, the less of a problem reentrance becomes. Increased passband bandwidth worsens the effects of reentrance because the edges of the multiple passbands approach each other. A deeper understanding of reentrance effects will develop as we consider specific filter structures in later chapters.

3.20   Coax

Coaxial cable is a two-conductor transmission line with a hollow outer conductor and a center conductor sharing the same longitudinal axis. The geometry is depicted in Figure 3-16. Coaxial cable is self-shielding which prevents radiation loss and coupling to adjacent lines. The center conductor is supported using a number of techniques including a solid dielectric, a

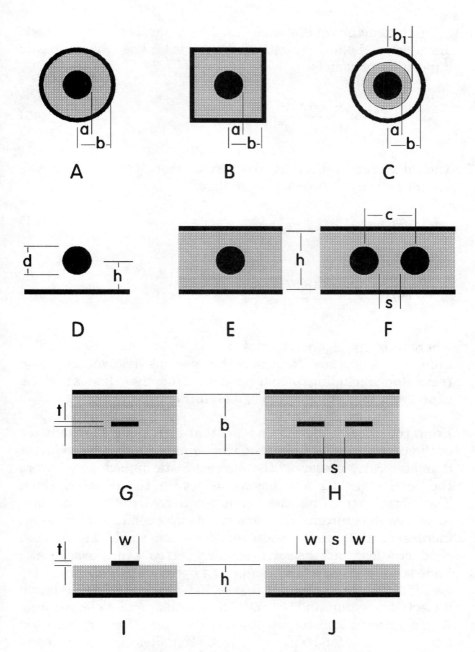

Figure 3-16 *Geometry of TEM and quasi-TEM mode transmission line structures.*

cellular structure of dielectric filled with gas, periodic dielectric disks or a helical spiral of dielectric. The characteristic impedance of coax is

$$Z_o = \frac{\eta_o}{2\pi\sqrt{\epsilon_r}} \ln\frac{b}{a} = \frac{60}{\sqrt{\epsilon_r}} \ln\frac{b}{a} \tag{40}$$

The attenuation of coax from conductor loss, α_c, and the attenuation from dielectric loss, α_d, are

$$\alpha_d(^{dB}/_m) = 91.207 \, f \, \sqrt{\epsilon_r} \, \tan\delta \tag{41}$$

$$\alpha_c(^{dB}/_m) = \frac{0.005694\rho_r\sqrt{f}}{z_o}\left(\frac{1}{a}+\frac{1}{b}\right) \tag{42}$$

where f is in gigahertz, and a and b are in meters. The unloaded Q versus frequency for various transmission-line resonators, including coaxial computed from these losses and the expression in section 3.15, is given in Figure 3-17.

From the above equations we see that increasing the diameter of the coax while retaining the same inner and outer conductor diameter ratio preserves the characteristic impedance, lowers the conductor loss and has no effect on the dielectric loss. Therefore, increasing the diameter decreases the loss and increases the unloaded Q. For a given outer conductor diameter, increasing the inner conductor diameter would reduce the conductor loss, but the corresponding decrease in characteristic impedance counters this trend. The optimum b/a to minimize loss is 3.59. The optimum b/a is independent of the relative dielectric constant and the ratio of conductor and dielectric loss. A b/a ratio of 3.59 corresponds to 76.7 ohms for air dielectric and 51.7 ohms for PTFE with a relative dielectric constant of 2.2. The impedance of strip-type transmission line structures which results in maximum unloaded Q is lower than the

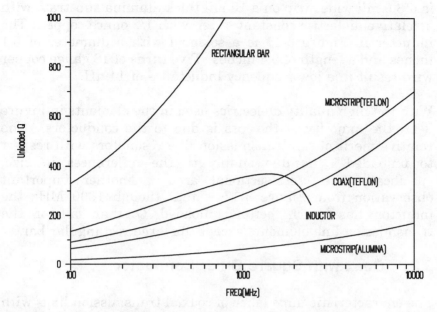

Figure 3-17 Unloaded Q vs. frequency for rectangular bar (thick stripline), microstrip on PTFE and alumina substrates and a solenoid inductor. See text for parameters.

optimum impedance for coax.

The coaxial line in Figure 3-17 is for copper semi-rigid coax with solid PTFE dielectric and an outer conductor diameter of 141 mils. This line is commonly used for UHF and microwave assemblies. The inside diameter of the outer conductor is 118 mils. The metalization roughness was estimated at 0.02 mils. The rectangular bar is a 500 mil wide and 100 mil thick strip between two ground planes separated by 500 mils with air dielectric. Thick stripline was used as the model for computing the loss and characteristic impedance which is 48.3 ohms. The roughness for this and the remaining lines is assumed to be 0.055 mils. The microstrip line on PTFE is a 152 mil wide strip ($\approx$50 ohms) on typical commercial 50 mil thick PTFE-glass microfiber with 1 ounce copper. The microstrip line on alumina is a 8.9 mil wide strip on a 10 mil thick alumina substrate with a relative dielectric constant of ten with 1/2 ounce copper. The inductor in Figure 3-17 is a solenoid with a diameter of 0.1 inches and a length of 0.4 inches. Five turns of 18 gauge copper wire result in a low-frequency inductance of 14 nH.

With the high quality dielectrics used in the elements in Figure 3-17, the majority of the loss is due to the conductors. The relative merit of the transmission line resonators with respect to unloaded Q is due simply to the differences in size, specifically the cross-sectional area. Another important observation from Figure 3-17 is that though 1600 MHz the inductor has much better unloaded Q than all of the transmission line elements except the large rectangular bar.

3.21 Coax with Square Outer Conductor

The characteristic impedance of coaxial transmission lines with a circular center conductor and a square outer conductor is reported by Cohn [11] as

$$Z_o = \frac{60}{\sqrt{\epsilon_r}} \ln \frac{1.0787b}{a} \tag{43}$$

which is reported to be within 1.5% for Z_o higher than 17 ohms and very accurate above 30 ohms. Reference [12] gives the relation

$$Z_o = \frac{138 \log \frac{b}{a} + 6.48 - 2.34A - 0.48B - 0.12C}{\sqrt{\epsilon_r}} \tag{44}$$

where

$$A = \frac{1 + 0.405\left(\dfrac{b}{a}\right)^{-4}}{1 - 0.405\left(\dfrac{b}{a}\right)^{-4}} \tag{45}$$

$$B = \frac{1 + 0.163\left(\dfrac{b}{a}\right)^{-8}}{1 - 0.163\left(\dfrac{b}{a}\right)^{-8}} \tag{46}$$

$$C = \frac{1 + 0.067\left(\dfrac{b}{a}\right)^{-12}}{1 - 0.067\left(\dfrac{b}{a}\right)^{-12}} \tag{47}$$

The accuracy of these expressions is undeclared in the reference.

3.22 Dielectric Loading

Physically short coaxial elements grounded at one or both ends are self-supporting and require no dielectric. A supporting dielectric material is used in physically longer coaxial lines to maintain the coaxial geometry. A material with a dielectric constant greater than one lowers the line characteristic impedance and lengthens the electrical length of the line for a given physical length, by reducing the propagation velocity by the square root of the relative dielectric constant.

The unloaded Q of coaxial elements is proportional to the diameter. Therefore, doubling the unloaded Q requires a four-fold increase in the cross-sectional area. The line length is determined by the desired electrical length and the operating frequency and is therefore not a selectable parameter. Loading the line element with a material with a dielectric constant greater than one reduces the volume required for a given unloaded Q, by reducing the required physical length by the square root of the dielectric constant.

Ceramic materials with excellent temperature stability and dielectric constant greater than 80 are now available which are suitable for RF and microwave applications. They provide significant foreshortening of transmission line elements. A representative list of commercial bulk dielectric and substrates is given in Table 3-7. Substrate materials from Arlon Microwave Materials Division are listed in Table 3-8.

A quarter-wavelength transmission line resonator with air dielectric is approximately 3.7 and 0.8 inches long at 800 and 3700 MHz, respectively. An 800 MHz filter requiring several resonators would be large. With material loading the resonators with a dielectric constant of 81, the length of the resonators are 0.41 and 0.09 inches at 800 and 3700 MHz, respectively. The dielectric material reduces the UHF resonator to a practical size, but the microwave resonator has become so small as to pose difficulties. The greatest utility of these high dielectric constant materials is therefore VHF through lower microwave frequencies where conventional lines are lengthy.

Resonators formed by metallic plating on coaxial dielectric material with a circular inner tube and rectangular or circular outer circumference are commercially available. Through the lower microwave frequency region, the loss tangent of the dielectric material is low and losses are primarily conductor losses. The unloaded Q is similar to conventional coaxial resonators of the same diameter, but the physical lengths are significantly reduced.

Table 3-7 *Commercial bulk dielectric and substrates materials. Manufacturers have additional materials and technical information.*

MATERIAL	ε_r	tol.($\pm$)	loss tangent 100 MHz	3 GHz	10 GHz
DIELECTRIC LABS					
CF	21.6	0.6		0.0003	
CB	29.0	0.6		0.0004	
CD	41.0	1.0		0.0004	
CG	67.5	2.0		0.0008	
NR	152	5.0		0.001	
MURATA					
Substrate P	21.4	1.0		0.00011	
Substrate H	38	1.0		0.00013	
Substrate K	91	3.0		0.001	
ROGERS					
RT/duroid 5880	2.20	0.02			0.0009
RT/duroid 5870	2.33	0.02			0.0012
ULTRALAM 2000	2.4-2.6	0.04			0.0019
RT/duroid 6002	2.94	0.04			0.0012
RT/duroid 6006	6.15	0.15			0.0019
RT/duroid 6010	10.2-10.8	0.25			0.0023
SIEMENS					
ε=21	20.9-22.4				0.00013
ε=29	28.4-29.3				0.0001
ε=38	37.4-38.8				0.00016
ε=88	89.8-91.8			0.00045	
TACONIC					
TLY-5	2.20	0.02			0.0009
TLY-3	2.33	0.02			0.0013
TLX-9	2.50	0.04			0.002
TLT-9	2.50	0.05	0.0009		
TRANS-TECH					
S8400	10.5				0.0001
8700	27.6-30.6	1.5		0.00005	0.0001
8500	35.7-36.4	1.5		0.00007	0.0002
8800	36.6-38.3	1.5		0.00008	0.0002
8600	80.0	1.5	0.0001	0.00025	

Table 3-8 *Arlon Microwave bulk dielectric and substrates materials.*

Substrate	ε_r	ε_r tol	ε_r TC	tan δ	Cost	Thick(mils)
Woven crossplied fiberglass reinforced PTFE						
CuClad 217	217 2.20	±0.02	−151	0.0009	MedHigh	5-250
CuClad 233	2.33	±0.02	−171	0.0012	MedHigh	5-250
CuClad 250	2.40 2.45 2.50 2.55 2.60	±0.04	−170	0.0019	Med	5-250
Woven non-crossplied fiberglass reinforced PTFE						
DiClad 880	2.17 2.20	±0.02	−160	0.0009	MedHigh	5-250
DiClad 870	2.33	±0.02	−161	0.0012	MedHigh	5-250
DiClad 522/527	2.4 2.45 2.50 2.55 2.60	±0.04	−153	0.0019	Med	5-250
Non-woven fiberglass reinforced PTFE						
IsoClad 917	2.17 2.20	±0.02	−157	0.0010	High	5-125
IsoClad 933	2.33	±0.02	−132	0.0013	High	5-125
Fiberglass reinforced and ceramic loaded PTFE						
CLTE	2.94	±0.04	<30	0.0025	High	5-250
AR320L	3.20	±0.05	N/A	0.0029	Low	15-250
AR350L	3.50	±0.15	−177	0.0026	Low	6-125
AR450L	4.50	±0.15	N/A	0.0026	Low	6-125
AR600	6.0	±0.15	N/A	0.0030	MedHigh	5-250
Ceramic loaded PTFE						
Epsilam 10	10.2	±0.25	−539	0.0027	High	25-100
DiClad 810	10.2 10.5 10.8	±0.25	−457	0.0020	VeryHigh	25-50

Notes:
1. (ε_r TC) is in parts per million per degree centigrade from 0 to 100 degrees centigrade.
2. The loss tangent is measured at 10 GHz.
3. Thick(mils) is the range of standard available substrate thicknesses.
4. The thermal coefficient of expansion of all materials in the X-Y direction is generally <40-50 ppm. The coefficient in the Z direction is about 200 for fiberglass substrates and <40 ppm for ceramic substrates.

3.23 Partial Dielectric Loading

The characteristic impedance of coaxial transmission line with partially coaxially-filled dielectric material is given by equation (40) with the relative dielectric constant replaced by an effective constant given by

$$\epsilon_e = \frac{\ln\dfrac{b}{a}}{\dfrac{1}{\epsilon_r}\ln\dfrac{b_1}{a} + \ln\dfrac{b}{b_1}} \tag{48}$$

Equation (48) is derived from Marcuvitz [13] and assumes radii are small with respect to a wavelength so that A from [13] is approximately equal to one.

3.24 Slabline

Slabline is a circular conductor centered between two infinitely-extending ground planes. The characteristic impedance is

$$Z_o = \frac{60}{\sqrt{\epsilon_r}}\ln\frac{4}{\pi d_n} \tag{49}$$

where d is the rod diameter normalized to the ground-to-ground spacing, h.

$$d_n = \frac{d}{h} \tag{50}$$

The above expression is accurate to about 1% for $d_n < 0.5$. A more accurate expression for a wider range of rod diameters, $0.1 < d_n < 0.9$, is

$$Z_o = \frac{59.9321}{\sqrt{\epsilon_r}} \ln \frac{4.00529}{\pi d_n} F_a \qquad (51)$$

where

$$F_a = 1 + e^{6.48695(d_n - 1.28032)} \qquad (52)$$

These expressions curve fit data obtained by numerical evaluation of the static capacitance of slabline by the moments method by Stracca [14].

3.25 Coupled Slabline

Coupled slabline is an important line structure because it is the basis of the popular round-rod combline and interdigital filter structures considered later. The previously cited Stracca reference include accurate numerical data as well as analytical expressions for coupled slabline. The numerical data is believed to be very accurate. The stated deviation from the numeric data from their analytical expressions is several percent, even for moderately wide rod spacings.

Given in Figure 3-18 is coupled slabline Z_m versus $c/2h$ for various d/h with a relative dielectric constant of one. c is the rod center-to-center spacing. Z_m is the arithmetic mean of the even and odd mode impedances.

$$Z_m = \frac{Z_{oe} + Z_{oo}}{2} \qquad (53)$$

The asymptotic value of Z_m with large $c/2h$ approaches the geometric mean impedance and the single slabline impedance, Z_o, given by equation (52).

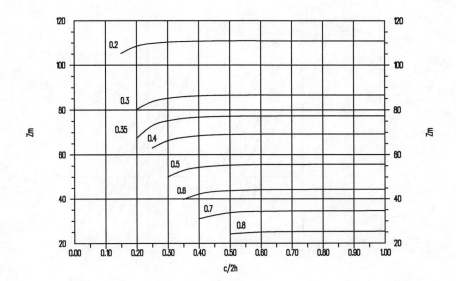

Figure 3-18 *Slabline arithmetic mean impedance versus c/2h with d/h as a parameter.*

The coupling coefficient, k, is related to the even and odd mode impedances by equations (34-38). k versus $c/2h$ as a function of d/h is given in Figure 3-19. k is plotted in Figure 3-19 with two scales. The coupling coefficient for the left scale ranges from 0.05 to 0.5 and is used with the left curves. This represents relatively tight coupling. The right scale ranges from 0.005 to 0.05 and is used with the right curves. This represents relatively loose coupling as might be experienced with narrowband filters.

3.26 Wire Over Ground

The characteristic impedance of a round wire over a ground plane is given by

$$Z_o = \frac{60}{\sqrt{\epsilon_r}} \ln \frac{4h}{d} \qquad (54)$$

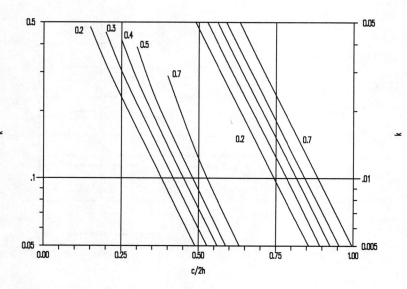

Figure 3-19 *Coupling coefficient, k, for coupled slabline versus c/2h with d/h as a parameter. The left scale is used for the left set of curves and the right scale is used with the right set of curves.*

for $d \ll h$.

3.27 Substrate Materials

Modern soft substrate laminates consist of metal foil (usually copper) laminated on both sides of a dielectric sheet which is a composite of various materials combined to produce the required mechanical and electrical performance parameters. Pure PTFE has a relative dielectric constant of 2.07 and an excellent loss tangent (dissipation factor) but poor mechanical stability. Electrical grade fiberglass has a dielectric constant of 6.0 and a higher loss tangent but adds mechanical strength. Substrates manufactured with PTFE and fiberglass have dielectric constants which typically range from 2.17 to 2.60. A PTFE content of 91% results in a dielectric constant of 2.2 and a content of 65% results in a dielectric constant of 2.6. PTFE material is more expensive than fiberglass. Therefore, substrates with a higher PTFE content have a lower dielectric constant, a lower loss tangent, poorer mechanical strength and greater cost.

Hard substrates are manufactured by depositing metalization on a fired ceramic substrate, providing a high dielectric constant and excellent loss tangent, but brittle substrates add significant circuit manufacturing cost. Ceramic powder is also used as a filler in PTFE based soft laminates, providing increased dielectric constants while retaining the advantages of soft substrates. Fiberglass fibers are oriented only in the foil plane which restricts expansion in the X-Y directions. These laminates are forced to expand in the Z direction (perpendicular to the printed circuit plane). Substrates with ceramic fillers have a lower temperature coefficient of expansion in the Z direction. A ceramic filler also has a higher thermal conductivity than fiberglass based substrates, resulting in improved heat-sinking properties. The hardness of ceramic fillers shorten manufacturing tool life.

Substrate materials from the Materials for Electronics Division of Arlon [15] are listed in Table 3-8. These substrates represent a family of substrate technologies which are suitable for a range of HF applications. Listed are the CuClad, DiClad, IsoClad and AR families.

Arlon CuClad laminate dielectrics consist of woven fiberglass and PTFE. As indicated in Table 3-8, they are available in a range of PTFE content percentages to achieve the desired performance trade offs. CuClad laminates are crossplied - each alternating ply of PTFE coated fiberglass woven-mat is rotated 90 degrees. This provides uniform electrical and mechanical properties in the X and Y directions (isotopy).

Arlon DiClad substrates are similar to CuClad except they are not crossplied. Each ply is oriented in the same direction This results in differences in the electrical and mechanical properties in the X and Y directions. The X-Y differential dielectric constant is approximately 0.01 to 0.015. This differential is smaller than the dielectric constant tolerance and should have little effect except with the narrowest of filter applications. In this event, attention to the orientation of the substrate is required during manufacture. DiClad substrates are available in sheets up to 36×72 inches, while crossplied CuClad is available in a maximum size of 36×36 inches.

Arlon IsoClad substrates are also a fiberglass/PTFE composite but the fiberglass is not woven. It consists of relatively long and randomly oriented fibers to improve the dimensional stability and dielectric constant uniformity. The random orientation of the fibers results in good isotopy. As with the woven laminates, a range of PTFE percentages are available to offer a selection of electrical performance and mechanical stability trade offs. The non-woven nature of IsoClad offers manufacturing advantages. During drilling, the woven fibers may be pushed through the hole instead of cleanly cut. Drill spindle RPM and feed rates are less critical.

Arlon CLTE is a ceramic filled PTFE based laminate. The nominal dielectric constant is 2.94 and the loss tangent is 0.0025 at 10 GHz, respectable for a substrate with a higher dielectric constant. The outstanding feature of CLTE is a maximum dielectric constant temperature coefficient of 30 parts per million per degree centigrade compared to several hundred parts per million for fiberglass reinforced PTFE substrates. The material composite in CLTE also results in excellent mechanical stability and a low thermal coefficient of expansion. Combined, these attributes produce a substrate with excellent electrical temperature stability. Reliability of plated through holes is also enhanced by the low coefficient of expansion which closely matches copper, reducing stress on the metalization. CLTE is therefore well suited for space applications.

Arlon AR series substrates are a fiberglass reinforced ceramic and PTFE composite. Ceramic fillers provide a significantly higher dielectric constant, even as high as popular hard ceramic substrates such as alumina. The selection of materials and manufacturing processes produce a substrate with electrical performance far superior to commercial thermoset fiberglass board such as G-10 and FR4, but with costs which are a fraction of conventional microwave substrates.

AR320L, AR350L and AR450L have dielectric constants of 3.2, 3.5 and 4.5, respectively. AR320L offers exceptionally tight dielectric constant tolerance for a lower cost material. AR450L has a relative dielectric constant very similar to G-10 and FR4 and therefore often can be used as a drop in replacement with lower loss and tighter specifications. Arlon AR600 with a nominal dielectric constant of 6.0 offers a higher dielectric constant for reduced circuit size at a moderate substrate cost.

Epsilam 10 and DiClad 810 are ceramic loaded PTFE without fiberglass reinforcement. Epsilam 10 has a nominal dielectric constant of 10.2. Unlike other members of the DiClad family, DiClad 810 is not fiberglass reinforced. It is available with nominal dielectric constants of 10.2, 10.5 or 10.8. While the cost of these higher dielectric constant substrates is greater, they offer a further reduction in circuit size.

3.28 Stripline

The characteristic impedance of stripline with a zero-thickness center strip is [16]

$$Z_o = \frac{\eta_o}{4\sqrt{\epsilon_r}} \frac{K(k)}{K(k')} \tag{55}$$

where

$$k = \mathrm{sech}\left(\frac{\pi w}{2b}\right) \tag{56}$$

$$k'=\tanh\left(\frac{\pi w}{2b}\right) \tag{57}$$

and $K(k)$ is the complete elliptic integral of the first kind. Hilberg [17] provides closed-form expressions for the ratio of the unprimed and primed elliptic integrals which are very precise. They are for $0 < k < 0.707$

$$\frac{K(k)}{K(k')}=\frac{\pi}{\ln\left(2\dfrac{1+\sqrt{k'}}{1-\sqrt{k'}}\right)} \tag{58}$$

and for $0.707 < k < 1$

$$\frac{K(k)}{K(k')}=\frac{1}{\pi}\ln\left(2\frac{1+\sqrt{k}}{1-\sqrt{k}}\right) \tag{59}$$

where

$$k'=\sqrt{1-k^2} \tag{60}$$

According to Wheeler [18], the characteristic impedance of stripline with a thick strip is

$$Z_0=\frac{30\ln}{\sqrt{\epsilon_r}}\left(1+\frac{c}{2}\left[c+\sqrt{c^2+6.27}\right]\right) \tag{61}$$

where

$$c = \frac{\dfrac{8}{\pi}}{\dfrac{W}{b-t} + \dfrac{\Delta W}{b-t}} \tag{62}$$

$$\frac{\Delta W}{b-t} = \frac{x}{\pi(1-x)} \left\{ 1 - \frac{1}{2} \ln \left[\left(\frac{x}{2-x} \right)^2 + \left(\frac{0.0796x}{W/b + 1.1x} \right)^m \right] \right\} \tag{63}$$

$$m = \frac{3}{1.5 + \dfrac{x}{1-x}} \tag{64}$$

$$x = \frac{t}{b} \tag{65}$$

The accuracy of these expressions is claimed to be better than 0.5% for $c > 0.25$.

Stripline conductor loss, λ_t, from Howe [19] is

$$\alpha_c(dB/m) = 0.0231 R_s \frac{\sqrt{\epsilon_r}}{Z_0} \frac{\partial Z_0}{\partial W} \left\{ 1 + \frac{16}{\pi c} - \frac{1}{\pi} \left[\frac{3x}{2-x} + \ln \left(\frac{x}{2-x} \right) \right] \right\} \tag{66}$$

where Z_o, c, and x were given earlier and

$$\frac{\partial Z_o}{\partial W} = \frac{30e^{-A}}{W'\sqrt{\epsilon_r}} \left[\frac{3.135}{Q} - c^2(1+Q) \right] \tag{67}$$

$$R_s = \sqrt{\pi \mu_o \rho f} \tag{68}$$

$$W' = \frac{8(b-t)}{\pi c} \tag{69}$$

$$A = \frac{Z_o \sqrt{\epsilon_r}}{30\pi} \tag{70}$$

$$Q = \sqrt{1 + \left(\frac{6.27}{c}\right)^2} \tag{71}$$

The dielectric loss is

$$\alpha_d(dB/m) = \frac{27.1 \sqrt{\epsilon_r} \tan\delta}{\lambda_o} \tag{72}$$

Stripline impedance versus the W/b ratio for various substrate dielectric constants is given in Figure 3-20. One ounce copper relative to $b=100$ mils is assumed.

3.29 Coupled Stripline

When two stripline conductors are placed side-by-side and share the ground planes, coupled stripline is formed. For zero strip thickness, the even and odd-mode impedances are

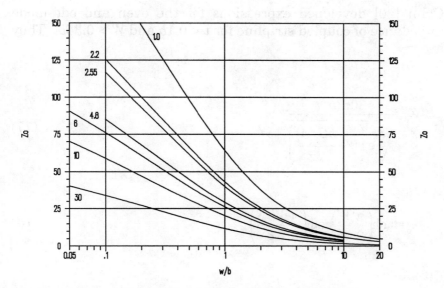

Figure 3-20 *Stripline impedance versus w/b for various substrate relative dielectric constants.*

$$Z_{oe} = \frac{30\pi}{\sqrt{\epsilon_r}} \frac{K(\acute{k}_e)}{K(k_e)} \tag{73}$$

$$Z_{oe} = \frac{30\pi}{\sqrt{\epsilon_r}} \frac{K(\acute{k}_o)}{K(k_o)} \tag{74}$$

where

$$k_e = \tanh\left(\frac{\pi}{2}\frac{W}{b}\right)\tanh\left(\frac{\pi}{2}\frac{W+S}{b}\right) \tag{75}$$

$$k_x = \sqrt{1-\acute{k}_x} \tag{76}$$

Cohn [20] developed expressions for the even and odd-mode impedance of coupled stripline for $t < 0.1b$ and $W > 0.35b$. They are

$$Z_{oe} = \frac{30\pi(b-t)}{\sqrt{\epsilon_r\left(W + \frac{bC_f A_e}{2\pi}\right)}} \tag{77}$$

$$k_o = \tanh\left(\frac{\pi}{2}\frac{W}{b}\right)\coth\left(\frac{\pi}{2}\frac{W+S}{b}\right) \tag{78}$$

$$Z_{oo} = \frac{30\pi(b-t)}{\sqrt{\epsilon_r\left(W + \frac{bC_f A_o}{2\pi}\right)}} \tag{79}$$

where

$$A_e = 1 + \frac{\ln(1+\tanh\theta)}{0.693} \tag{80}$$

$$A_o = 1 + \frac{\ln(1+\coth\theta)}{0.693} \tag{81}$$

$$\theta = \frac{\pi S}{2b} \tag{82}$$

$$C_f = 2\ln\left(\frac{2b-t}{b-t}\right) - \frac{t}{b}\ln\left[\frac{t(2b-t)}{(b-t)^2}\right] \tag{83}$$

3.30 Microstrip

With microstrip, a portion of the electric fields are in the dielectric between the strip and the ground plane while other fields exist in the region above the strip with air as a dielectric. At frequencies where the electrical distance in the dielectric material between the strip and ground plane is much less than a wavelength, microstrip behaves as a non-dispersive *TEM* line. Unlike moding in pure *TEM* lines, the transition in microstrip from *TEM* to quasi-*TEM* is not sudden. With increasing frequency, as the substrate thickness of microstrip becomes appreciable, the propagation velocity and the characteristic impedance of the line increase.

For pure *TEM* mode transmission lines, there are no longitudinal components of the fields, and the various definitions of the characteristic impedance, such as voltage over current, the voltage squared over power, etc., yield identical results. When longitudinal fields exist, the values of the voltage and current are reference-point dependent. Nevertheless, expressions for microstrip characteristic impedance which is a useful circuit parameter have been developed.   An excellent review of quasi-*TEM* impedance definitions is given by Hoffman [21]

We begin with the low-frequency static characteristic impedance as a function of strip width normalized to the substrate height, a substrate dielectric constant of one and zero strip thickness, Z_o

$(W/h, f = 0, \varepsilon_r = 1, t = 0)$. To avoid lengthy expressions, we will shorten this expression to Z_o (0) to designate only the static constraint and the status of other parameters, such as thickness, is defined as we proceed.

A number of workers have contributed to the determination of the characteristic impedance of microstrip. An early and accurate benchmark contribution was made by Bryant and Weiss [22] who found the characteristic impedance of microstrip using numeric techniques. The program MSTRIP [23] which came out of this work served as a standard for some time. Unfortunately this work addresses the static case only and the numeric approach is too slow for interactive circuit simulation programs which are heavily used for circuit design today. Much of the following work was an attempt to find closed form analytical expressions to approximate the earlier and accurate numeric data, and to add the effects of dispersion to increase accuracy at higher frequencies.

For the static case, with zero strip thickness, the analytic expressions of Hammerstad and Jensen, are widely regarded as among the best. They are

$$Z_o(0) = 60\ln\left(\frac{F_1 h}{W} + \sqrt{1 + (2h/W)^2}\right) \tag{84}$$

where

$$F_1 = 6 + (2\pi - 6)e^{-(30.666h/W)^{0.7528}} \tag{85}$$

The above expressions are within 0.01% for $W/h < 1$ and 0.04% for $W/h < 1000$. The characteristic impedance with $\varepsilon_r > 1$ is then found from

$$Z_o(\epsilon_r) = \frac{Z_o(\epsilon_r = 1)}{\sqrt{\epsilon_{r,eff}}} \tag{86}$$

where $\epsilon_{r,eff}$ is the effective dielectric constant. For microstrip, the effective dielectric constant is less than substrate material dielectric constant because a portion of the fields are in air. The static value of the effective dielectric constant is given by

$$\epsilon_{r,eff} = \frac{\epsilon_r + 1}{2} + \frac{\epsilon_r - 1}{2}\left(1 + \frac{10h}{W}\right) - ab \tag{87}$$

where

$$a = 1 + \frac{1}{49}\ln\left[\frac{\left(\frac{w}{h}\right)^4 + \left(\frac{w}{52h}\right)^2}{\left(\frac{w}{h}\right)^4 + 0.432}\right] + \frac{1}{18.7}\ln\left[1 + \left(\frac{w}{18.1h}\right)^3\right] \tag{88}$$

$$b = 0.564\left(\frac{\epsilon_r - 0.9}{\epsilon_r + 3.0}\right)^{0.053} \tag{89}$$

Static microstrip impedance versus w/h for various relative dielectric constants is given in Figure 3-21. The static impedance is accurate when the substrate is sufficiently thin for a given operating frequency that line dispersion is nil and the impedance is not a function of frequency. One ounce copper is assumed. Heavier copper lowers the impedance and lighter copper increases the impedance. These effects are minimal for wider lines and are typically only significant for high line impedance. See also Figure A-3.

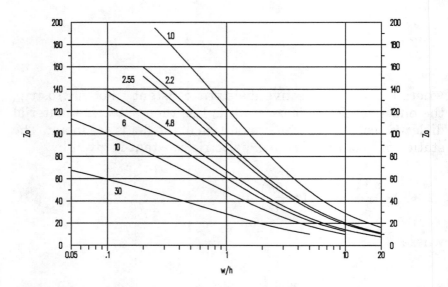

Figure 3-21 *Static microstrip impedance versus w / h for various substrate relative dielectric constants.*

Given in Figure 3-22 is the non-static characteristic impedance of microstrip for various relative dielectric constants. The strip width for each relative dielectric constant is adjusted for a static impedance of 50 ohms. Because dispersion in microstrip is a result of non-homogenous media, it is not surprising that the degree of dispersion increases with increasing substrate relative dielectric constant and vanishes as the substrate dielectric constant approaches the dielectric constant of the media above the strip (air). Dispersion is proportional to the substrate thickness which is 1 millimeter in Figure 3-22. For a substrate thickness of one-half millimeter the frequency scale in Figure 3-22 is doubled.

With a substrate relative dielectric constant of 10 and a thickness of 25 mils, dispersion increases the impedance 2% over the static value at approximately 12.6 GHz. For a substrate relative dielectric constant of 2.55 and a thickness of 50 mils, the impedance is increased 2% at approximately 9.8 GHz.

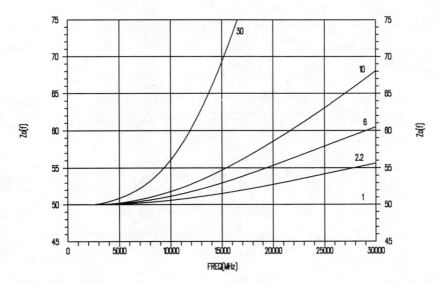

Figure 3-22 *Microstrip Z_o vs. frequency for various substrate dielectric constants. Each line width is adjusted for 50 ohm static Z_o. The substrate thickness is 1 mm.*

3.31 Coupled Microstrip

Given in Figure 3-23 (left vertical scale and flat curves) are the required microstrip $W=w/h$ versus coupling coefficient which result in a static geometric mean impedance of 50 ohms for various substrate relative dielectric constants. For loose coupling, the curves are flat and correspond to the isolated (single-line) case and the impedance asymtotically approaches the impedance given by equation (86). Also given in Figure 3-23 (right vertical scale and down-sloping curves) are $S=s/h$ versus coupling coefficient for various substrate relative dielectric constants. Notice the spacing is a strong function of the required coupling but only a moderate function of the relative dielectric constant. The relationship of the even and odd mode impedances, the geometric mean impedance Z_o and the coupling coefficient, k, are given by equations (34) through (38).

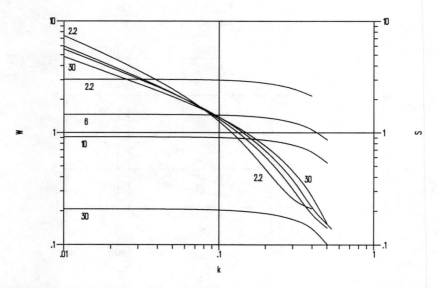

Figure 3-23 *Static coupled microstrip w/h for Z_o of 50 ohms versus coupling coefficient, k, for various relative dielectric constants (left scale and flat curves) and s/h versus coupling coefficient (right scale and sloping curves).*

3.32 Stepped-Impedance Resonators

The distributed resonators we have previously envisioned consist of a resonant line of uniform impedance. Later we will investigate loading a transmission line with a lumped or distributed reactance to resonate a line which is less than a self-resonant length. In this section we investigate the properties of distributed resonators formed by cascading two lines with a different characteristic impedance. Electrical resonance is achieved with a shorter physical length. Such a stepped-impedance resonator is depicted in coaxial form in Figure 3-24.

A conventional quarter-wavelength transmission-line resonator consists of continuously distributed inductance and capacitance. At the grounded, low voltage end of the resonator, the capacitance to ground has minimal effect and at the open high

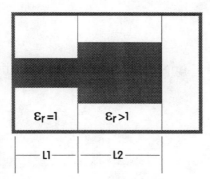

Figure 3-24 *Stepped-impedance resonator consisting of a high-impedance grounded section and a lower impedance terminating section.*

impedance end of the resonator, the inductance has minimal effect. The distributed inductance of a line is increased in relation to the capacitance by increasing the line impedance. Therefore, by increasing the impedance of the grounded end of a resonator and decreasing the impedance of the open end of a resonator, both the effective inductance and capacitance are increased which lowers the resonant frequency. It is therefore possible to decrease the resonant frequency for a given physical length and to improve stopband performance by using this stepped-impedance approach.

We define K, the impedance ratio, as [24]

$$K = \frac{Z_2}{Z_1} \tag{90}$$

Resonance is achieved when

$$\tan(\theta_1)\tan(\theta_2) - K = 0 \tag{91}$$

We normalize the lengths to a quarter-wavelength in free space.

$$L_x = \frac{l_x}{\lambda/4} \tag{92}$$

The total physical length is the sum of the individual section lengths.

$$L_t = L_1 + L_2 \tag{93}$$

Then the total resonator length is

$$L_t = \frac{2}{\pi\sqrt{\epsilon_r}} \left[\tan^{-1}\left(\frac{H}{1-K}\right) + \frac{\pi L_1}{2}\left(\sqrt{\epsilon_r}-1\right) \right] \tag{94}$$

where

$$H = \tan\left(\frac{\pi L_1}{2}\right) + \frac{K}{\tan\left(\frac{\pi L_1}{2}\right)} \tag{95}$$

For a given impedance ratio, K, and dielectric constant, ϵ_r, for section L_2, there is a minimum overall resonator length which is achieved when

$$L_{1min} = \frac{2}{\pi}\tan^{-1}\left\{ \left[\frac{K(1-\sqrt{\epsilon_r}K)}{\sqrt{\epsilon_r}-K}\right]^{1/2} \right\} \tag{96}$$

The overall resonator length is plotted in Figure 3-25 for various stepped-impedance resonator configurations. Curves a and b are with $\epsilon_r = 1$ in both the high and low-impedance sections. Curve a is with $K=0.5$ and curve b is with $K=0.2$. Notice that even with air as a dielectric, with $K=0.2$ the physical length of the stepped-impedance resonator is only 60% of the length of a

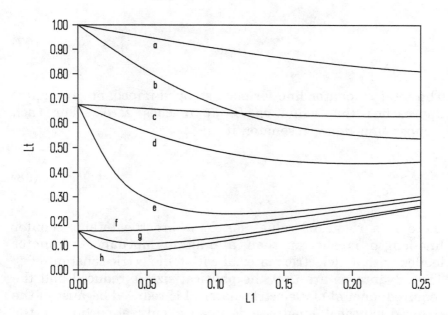

Figure 3-25 *Total length of a stepped-impedance resonator relative to a quarter-wavelength versus L_1 for various steeped-impedance configurations. See text for details.*

conventional quarter-wavelength resonator. Curves c, d and e are with $\varepsilon_r = 2.2$ in the low-impedance section and impedance ratios of 0.5, 0.2 and 0.05 respectively. The physical length for case e is only 25% of the length of conventional air-dielectric resonator when L_1 is the optimum length. Curves f, g and h are with $\varepsilon_r = 38.6$ in the low-impedance section with impedance ratios of 0.16, 0.05 and 0.02 respectively.

For a conventional constant impedance dielectric loaded resonator, the line-length normalized to a quarter-wavelength resonator in air is

$$L_{tu} = \frac{1}{\sqrt{\epsilon_r}} \tag{97}$$

The total resonator line for each stepped-impedance resonator approaches this value as L_1 approaches zero. For each stepped-impedance resonator, if

$$K < \frac{1}{\sqrt{\epsilon_r}} \tag{98}$$

there is a range of values for L_1 where the total resonator line-length is shorter than a uniform-impedance resonator loaded with a dielectric material with a dielectric constant of ϵ_r. The advantages are that the physical size is reduced and the required amount of dielectric material is reduced because of the reduced physical length and because only a portion of the resonator includes dielectric material.

Another advantage of dielectric resonators is that the reentrance mode is increased in frequency. The frequency ratio number, N, which is three for conventional quarter-wavelength resonators, is found by solving the expression

$$\frac{\tan\left(\dfrac{\pi L_1 N}{2}\right)\tan\left(\dfrac{\pi\sqrt{\epsilon_r}L_2 N}{2}\right) - K}{\tan\left(\dfrac{\pi L_1 N}{2}\right) + K\,\tan\left(\dfrac{\pi\sqrt{\epsilon_r}L_2 N}{2}\right)} = 0 \tag{99}$$

Shown in Figure 3-26 are values of N versus L_1 with K as a running parameter. We can see that case e in Figure 3-25, which has a four to one reduction in the resonator length, has a first reentrant mode frequency which is ten times the fundamental resonant mode instead of three times the fundamental mode which it would be for a conventional

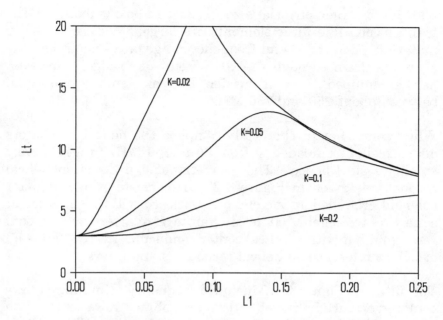

Figure 3-26 *Stepped-impedance resonator frequency ratio, N, versus L_1 with various impedance ratios, K.*

resonator. These curves are independent of ε_r.

The unloaded Q of stepped-impedance resonators is less than that of conventional quarter-wavelength resonators. Unloaded Q for stepped-impedance resonators is discussed by Stracca and Panzeri [25]. With low-loss dielectric material loading section two of the stepped-impedance resonator, the unloaded Q is degraded less than 25% with foreshortening less than 50%. The unloaded Q degrades rapidly for stepped-impedance resonators less than 25% of the length of conventional resonators.

3.33 Helical Resonators

We have seen that the self-capacitance of lumped inductors forces a small physical size at high frequencies and limits the available unloaded Q. Transmission line elements incorporate the distributed capacitance as an integral part of the structure, allowing a larger physical size. However, below about 1 GHz, transmission line filter elements are large. Modern dielectric materials push the useful frequency range of transmission line elements down through the UHF frequency range. However, neither lumped or transmission line elements are optimal between about 250 and 750 MHz.

While somewhat mechanically complex, the helical resonators offer excellent unloaded Q from 30 to 750 MHz in a relatively small physical space. The geometry of a five-section helical resonator is given in Figure 3-27. It consists of a conductor solenoid grounded at one end and enclosed within a circular or square outer shield. It is similar to a quarter-wave coaxial resonator, but with a helical center conductor. Zverev [26] is an excellent reference on helical resonators and filters.

The filter in Figure 3-27 includes frequency trimming screws centered on each helix and internal coupling screws centered on wall openings at the top of the resonator cavities. Electrostatic coupling through these wall openings is reduced as the screws penetrate the housing. External coupling is provided by capacitive probes at the top of the end resonators. Coupling is increased with larger disks or placing the disks closer to the helix.

Currents are highest at the grounded end of the solenoid and electric fields are highest at the floating end of the solenoid. The solenoid may be returned to ground at the bottom of the shield as opposed to the side, provided the seam between the side and bottom has a low resistance to minimize losses. Dielectric losses are greatest at the top or floating end of the solenoid. The ends of the shield should extend beyond the ends of the solenoid by one-quarter the shield diameter. They may be either open or closed, but a closed shield eliminates external fields.

Because the helix is supported at one end and is effectively a long spring, it is susceptible to mechanical vibration which

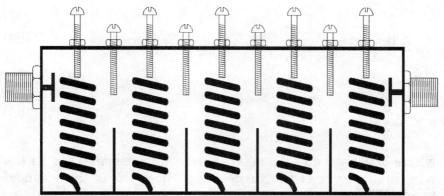

Figure 3-27 *Example helical resonator bandpass filter structure.*

modulates the center frequency and phase response of the resonator. To avoid this problem, the solenoid may be wound on a dielectric support form. A tubular form with a low dielectric constant and low loss tangent has a minimal effect.

If a square shield is used, D in the following expressions should be replaced with $1.2\ S$, where S is the length of the sides of the shield. The highest unloaded Q for helical resonators is obtained over a very narrow range of geometric parameters.

For example, the diameter of the solenoid should be 45 to 60% of the shield diameter. The unloaded Q is degraded by 33% if the solenoid to shield diameter ratio is 25%. The following design parameters for helical resonators assume the following geometry

$$0.45 < \frac{d}{D} < 0.6 \tag{100}$$

$$\frac{b}{d} = 1.5 \tag{101}$$

$$0.4 < \frac{d_o}{\tau} < 0.6 \tag{102}$$

$$\tau < \frac{d}{2} \tag{103}$$

where d is the helix diameter (wire center-to-center), d_o is the wire diameter, τ is the turns per inch and b is the solenoid length Then

$$L(\mu H\ per\,axial\ inch) = 0.025 \left(\frac{Nd}{b}\right)^2 \left[1 - \left(\frac{d}{D}\right)^2\right] \tag{104}$$

$$C(pF\ per\ axial\ inch) = \frac{0.75}{\log\dfrac{D}{d}} \tag{105}$$

where N is the total number of turns. The characteristic impedance and unloaded Q of the resonator are then

$$Z_o = \frac{183Nd}{b} \sqrt{\left(1 - \frac{d}{D}\right)^2 \log\frac{D}{d}} \tag{106}$$

$$Q_u = \frac{50D\sqrt{f_o}}{\rho_r} \tag{107}$$

where D is in inches and f_o is in megahertz. The length of the solenoid in inches for resonance after a 6% reduction due to fringing effects is

$$b = \frac{235}{f_o\sqrt{LC}} \tag{108}$$

Assuming the resonant frequency is known and a certain unloaded Q is required, the design proceeds as follows

$$D = \frac{Q_u \rho_r}{50\sqrt{f_o}} \tag{109}$$

$$N = \frac{1900}{f_o D} \tag{110}$$

The remaining physical parameters are derived from the geometric relationships presented earlier. If the total number of turns, N, is less than three, the expressions are not valid and a smaller unloaded Q should be selected. This is an indication that a conventional distributed line instead of a helical line should be considered.

The primary difficulties with helical resonators are determining and controlling the coupling between resonators and external coupling to the terminations. The five-section helical filter in Figure 3-27 uses capacitive (electrostatic) coupling at the high-potential end (floating) of the end resonators. Depicted in Figure 3-28 are two forms of magnetic external coupling near the grounded end of the end resonators; a wire loop and a direct tap of the helix. If the loop extends from the center pin to the body of the connector inside the housing, the connector may be rotated to adjust the coupling. Two forms of internal coupling are depicted in Figure 3-28: a hole in the housing wall between the first and second resonator and a transfer loop from the second to third resonator. The first method is more economic for high volume production while the latter method is more readily adjusted and is suitable for small quantity runs where the effort

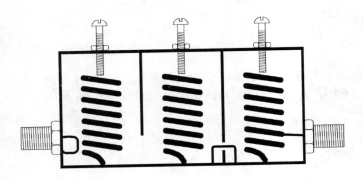

Figure 3-28 *Three-section helical bandpass with a mixture of external and internal coupling methods.*

to precisely determine the required hole size is unjustified.

The Zverev reference gives a plot and expression for the resonator-to-resonator coupling for a variable size magnetic-hole in the shielding wall. A simple electrical measurement is performed which determines the resonator-to-resonator coupling. The measurement is only required for a few hole sizes and a smooth curve is drawn through the data points to develop a curve for coupling versus hole size. Unfortunately, the test cases are only valid for each unique geometry and center frequency. However, this is a powerful and general method for design of practical filters which refuses to yield to more mathematical approaches.

3.34 Dielectric Resonators

Microwave resonators may be formed using bulk ceramic material, typically in cylindrical form. The resonant frequency of a bulk resonator well isolated from an enclosing housing is given [27] to within about 2% by

$$f_r = \frac{34}{a\epsilon_r}\left(\frac{a}{H}+3.45\right) \tag{111}$$

for

$$0.5 < \frac{a}{H} < 2.0 \tag{112}$$

$$30 < \epsilon_r < 50 \tag{113}$$

where a and H are the cylindrical radius and height in millimeters. Algorithms for the resonant frequency in proximity to a housing and for more general parameters are discussed by Kajfez.

3.35 Waveguide

At higher microwave frequencies, TE or TM mode waveguide can be used as resonators. High unloaded Q is achieved at the expense of large physical size for the filter. Typical bandpass structures consist of propagating sections of waveguide acting as half-wavelength resonators which are coupled via discontinuities in the guide such as vanes, irises and posts. Waveguide filter structures and design procedures are given in Matthaei, et. al. [28]. Characteristics of discontinuities in waveguide are given throughout Marcuvitz [13].

The unloaded Q of air-filled copper waveguide is

$$Q_u = \frac{A}{a\sqrt{f}} \tag{114}$$

where A is a constant, a is the broad dimension of rectangular guide or the diameter of circular guide and f is the operating frequency in GHz. For TE_{mo} modes in rectangular guide with a b/a ratio of 0.45, $A = 3500$ at an operating frequency of 1.3 times the waveguide cutoff frequency and $A = 4500$ at 2 times the cut off. For rectangular guide with a b/a ratio of 0.5, $A = 3800$ at 1.3 times the cutoff and 4800 at 2 times the cutoff. For the TE_{11} mode in circular guide, $A = 6000$ at 1.3 times the cutoff and 9000 at 2 times the cutoff.

3.36   Evanescent Mode Waveguide

Waveguide operated at frequencies below the dominant mode cutoff does not propagate. In fact, well below the cutoff, the attenuation is frequency independent and is only a function of the guide length and the cross-sectional dimensions. This phenomena serves as the basis of mechanically variable attenuators. The fact that waves below cutoff do not propagate but die exponentially leads to the term evanescent.

In 1950, Barlow and Cullen [29] reported that a resonance is formed when evanescent mode guide is terminated with a capacitive reactance. Edson proposed the use of evanescent mode waveguide for microwave filters [30] and Craven [31,32] published useful design principles. As it turns out, evanescent mode waveguide behaves as a tee or pi of frequency dependent, lumped inductors. From Craven, the characteristic impedance of evanescent guide is

$$Z_o = jX_o + R \tag{115}$$

For the lossless case

$$Z_o = jX_o \tag{116}$$

where

$$X_o = \frac{120\pi b}{a\sqrt{\left(\frac{\lambda}{\lambda_c}\right)^2 - 1}}$$

(117)

The "propagation" constant, γ, is given by

$$\gamma = \frac{2\pi}{\lambda}\sqrt{\left(\frac{\lambda}{\lambda_c}\right)^2 - 1}$$

(118)

where λ is the free-space wavelength, λ_c is the guide cutoff wavelength, a is the guide wide dimension (width) and b is the guide narrow dimension (height). The reactance of the lumped inductors in the tee model for evanescent mode waveguide given in Figure 3-29 are

$$XL_a = XL_c = X_o \tanh\frac{\gamma l}{2}$$

(119)

$$XL_b = \frac{X_o}{\sinh\frac{\gamma l}{2}}$$

(120)

and the reactance for the pi model are

$$XL_1 = XL_2 = X_o \coth\frac{\gamma l}{2}$$

(121)

$$XL_3 = X_o \sinh\frac{\gamma l}{2}$$

(122)

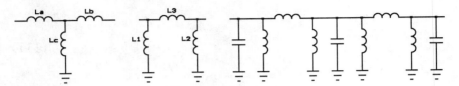

Figure 3-29 *Tee and pi lumped inductor models for evanescent mode waveguide and a filter created by loading the guide with capacitive elements.*

where l is the length of the evanescent mode waveguide section in the same units as γ. Because γ is a function of frequency, the inductors are a function of frequency. For bandpass filters to about 20% bandwidth, the center frequency is specified and the inductance variation has minimal effect. Snyder [33] gives design procedures suitable for wider bandwidth.

As is shown in Figure 3-29, a bandpass filter is formed by loading the evanescent mode waveguide with capacitive elements. Two types capacitive elements are typically used. A short section of guide loaded with dielectric material forms a capacitive element. This is often used for wideband evanescent mode filters. For narrowband filters, a post perpendicular to the broad dimension and which approaches the opposite wall forms a parallel plate capacitor. A typical post diameter is one tenth the broad dimension of the waveguide. An estimate of the capacitance is given by equation (29). The effects of fringing capacitance and the inductance of the post is compensated for by tuning the post length. Caution should be exercised in selecting suitable capacitive elements because discontinuities in waveguide below cutoff behave differently than discontinuities in propagating guide.

It is also important to recognize the strong lumped-element nature of evanescent mode waveguide. The element values in evanescent mode filters are found using strictly lumped-element theory with the required evanescent mode waveguide section lengths found using equations (119-122). The lumped-model nature of these filters is so complete that lumped-element piston trimmer capacitors may be used for the capacitive elements. The design of evanescent mode bandpass filters is considered in Chapter 8.

3.37 Evanescent Mode Unloaded Q

The unloaded Q of evanescent rectangular waveguide given by Craven and Mok [31] is

$$Q_u = \frac{\omega\mu}{R_s}\frac{ab}{2}\frac{1-\frac{1}{2}\left(\frac{f}{f_c}\right)^2}{a\left[1-\frac{1}{2}\left(\frac{f}{f_c}\right)^2\right]+b} \tag{123}$$

where a is the waveguide width in inches, b is the waveguide height in inches, f_c is the waveguide cutoff frequency, f and ω are the operating frequency and radian frequency, R_s is the sheet resistivity and μ is the permeability. The sheet resistivity is

$$R_s = \sqrt{\pi f \mu \rho} \tag{124}$$

which is 2.609×10^{-6} ohms per square for copper at 1 Hz. ρ is the absolute conductor resistivity. A plot of equation (123) for copper WR90 X-band waveguide (a=900 mils, b=400 mils) is given in Figure 3-30. Also included is the unloaded Q for propagating waveguide above the cutoff (6557 MHz). While $b<a$ for standard rectangular waveguide, square guide provides a proportionally higher unloaded Q for a given width.

Circular waveguide, or tubing, also offers a high unloaded Q. Q_u of circular evanescent mode waveguide is reported by Snyder [32] to be

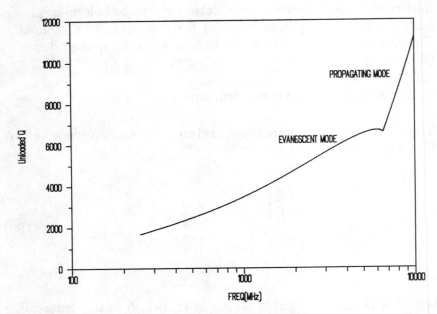

Figure 3-30 *Unloaded Q of copper WR90 rectangular waveguide operated in the evanescent mode (below cutoff) and above cutoff. The inside waveguide dimensions are a=900 mils and b=400 mils.*

$$Q_u = \frac{\pi f \mu d}{R_s} \left[\frac{0.404 + 0.405\left(1 - \frac{f^2}{f_c^2}\right)}{1.144 + 0.338\left(1 - \frac{f^2}{f_c^2}\right)} \right] \tag{125}$$

The unloaded Q may be substantially reduced by losses in the loading capacitance. If the capacitance is realized as a simple machine screw, the threads exposed within the waveguide should be removed and the remaining smooth metal should be plated with a highly conductive material such as copper, silver or gold. A low resistance contact should be insured at the grounded end of the tuning screw. Even with these precautions, the unloaded Q of evanescent mode waveguide filters with capacitive posts is reported by several workers to be no more than 40 to 70% of the theoretical evanescent mode guide unloaded Q. A narrowband 1 GHz filter constructed in 875 mil square aluminum tubing required several picofarads of loading capacitance. The capacitance was realized with commercial piston trimmers because the spacing for a simple post would be exceptionally close. With piston trimmers, the resonator unloaded Q as estimated from the filter insertion loss was only 200.

3.38 Superconductors

Superconductors have a resistance of zero ohms at dc and avoid magnetic field penetration [34]. Certain materials become superconducting when the temperature drops below a critical value (4 degrees Kelvin for mercury and 7 degrees for lead). This phenomena has been known to exist in common conductors such as lead and mercury since 1911. Achieving these temperatures requires the use of expensive refrigerator equipment or liquid helium which is inefficient as a coolant. Today materials are available with critical temperatures as high as 123 degrees Kelvin. Superconductors with critical temperatures above 77 degrees Kelvin are especially practical because liquid nitrogen can be used as a coolant.

Many conventional design precepts are invalid for superconducting materials. While the resistance of superconductors at dc is zero, the resistance increases with increasing frequency. Conventional metallic loss increases with the square root of frequency while supercondutor losses increase with the frequency squared. At frequencies higher than 100 GHz, the resistance of superconductors and conventional copper converge. Nevertheless, though microwave frequencies, the conductivity, and therefore the unloaded Q, of resonators constructed with superconductive materials is substantially higher than with conventional metalizations. Also, because superconductors expel magnetic fields, field penetration levels do not follow conventional skin effect laws. The penetration depth is a function of the superconducting material, but it is typically less than the conventional skin depth and thin superconducting films are practical. It is also necessary to keep the current density and magnetic field strength below critical levels or superconductivity is lost.

Given in Figure 3-31 are stripline resonator unloaded Q for superconducting materials and copper on 25 mil thick substrates. Increasing the substrate thickness increases the unloaded Q when the Q is conductor limited as with conventional copper. For superconduction on magnesium oxide, the Q is dielectric limited, and increasing the substrate thickness offers limited benefit. Dielectric unloaded Q is equal to the inverse of the material loss tangent. Superconducting planar microwave filter structures have been demonstrated with unloaded Qs approaching that of much larger waveguide. Because the dielectric Q of sapphire is very high, future use of this support material for superconductors will provide very high resonator Q. To realize the full unloaded Q potential of superconducting filters, it is imperative that radiation problems be carefully managed.

3.39 Material Technology Unloaded Q Summary

A number of technologies are available for the implementation of the reactive elements in the filter structure. Early implementations included lumped inductors and capacitors. Bulk quartz crystal piezoelectric resonators, coaxial and waveguide elements matured during and after WWII. Later, planar structures such as stripline and microstrip were developed. More recent advancements include high-dielectric constant materials and MMIC structures. Each have application-specific advantages. Figure 3-32 diagrams unloaded Q versus frequency for a number of component technologies. This data should be considered as only a guideline; precise

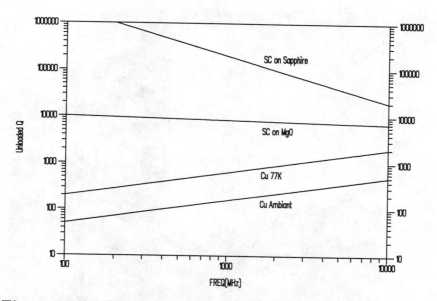

Figure 3-31 *Stripline resonator unloaded Q vs. frequency for conventional copper and superconducting material technologies.*

values are a function of many parameters and tradeoffs. In general, the upper left limit of component regions are defined by physical size. These limits are extended by a larger than usual size. With lumped inductors, the lower right region represents small physical size. In the extreme, wire wound inductors are abandoned and process inductors are used. The upper right limits are defined by parasitics or moding and are more difficult to extend.

For process technologies other than crystal, filters are readily constructed as long as the filter loaded Q is much less than the unloaded Q of the component. Therefore the shaded regions extend downward indefinitely. In other words, the shaded regions limit only how narrow a bandpass filter can be constructed. An exception is bulk piezoelectric crystals where the static capacitance limits how wide a bandwidth is feasible. The crystal shaded region cannot be extended downward. This creates a large gap in the chart. Below the practical frequency

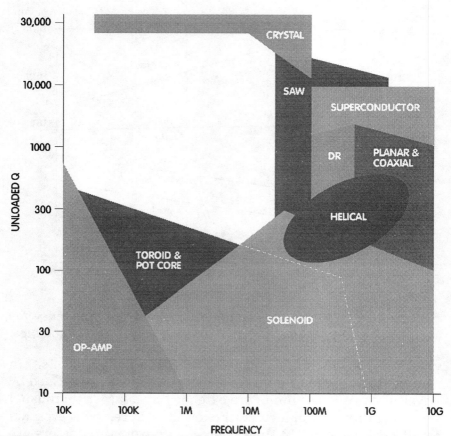

Figure 3-32 *Unloaded Q vs. frequency for various technologies. Some region limits are defined by practical issues such as size and are extendable while others are defined by rigid barriers such as moding.*

range of SAW resonators, bandwidths greater than about 0.1% and less than about 2% are very difficult to achieve. In general, system design should avoid these filtering requirements. Less common technologies capable of bridging this gap include ceramic piezoelectric resonators and mechanical/acoustical resonators.

3.40 Unloaded Q Versus Volume

Given in Figure 3-33 are the required volumes to achieve a given unloaded Q for various reactor/resonator technologies. The metalization in each case is copper. The unloaded Q includes both metalization and dielectric losses where appropriate but has no safety margin. Radiation is ignored.

The required solenoid diameter includes spacing from adjacent components of a solenoid diameter on all sides and two diameters at each end. The solenoid inductor diameter is adjusted to achieve the required unloaded Q. The length to diameter ratio is fixed at 4:1 and the wire gauge is selected to achieve optimum spacing for best unloaded Q. The inductance naturally increases with increasing solenoid diameter. The solenoid inductor curve terminates near an unloaded Q of 500 because parallel resonance associated with parasitic capacitance degrades the unloaded Q with increasing diameter. In fact, because the solenoid inductance and Q are heavily dependent on the poorly predictable parasitic capacitance, smaller solenoid diameter is advised. An inductor curve for 3000 MHz is absent because the volume is so small it is off the volume scale at the bottom.

All transmission line resonators in the chart are designed with a characteristic impedance of 50 ohms. The coaxial resonator curves assume solid PTFE dielectric. Coaxial resonators are self shielding so additional volume to avoid proximity to other components is unnecessary. The coax outer conductor diameter is adjusted to achieve the unloaded Q. The diameter ranges from 50 to 1300 mils at 300 MHz and from 50 to 500 mils at 3000 MHz.

Microstrip includes a margin on each side of the strip and a housing cover separation equal to five times the substrate thickness. The microstrip lines on PTFE and G-10 assume 1 ounce copper and a substrate roughness of 0.075 mils. The microstrip substrate thickness is adjusted to provide the

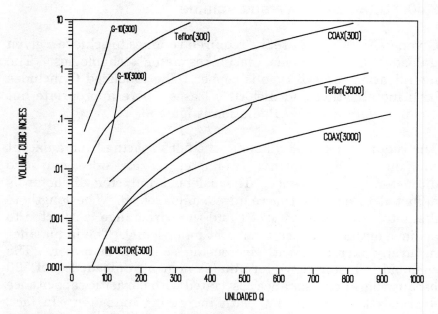

Figure 3-33 *Required volume vs. unloaded Q for inductor, coax, microstrip on PTFE and microstrip on G-10 (FR4) at 300 and 3000 MHz. Inductor volume at 3000 MHz is off scale at the bottom.*

required unloaded Q and the strip width is adjusted to achieve 50 ohms. The thickness of the substrate ranges from 10 to 125 mils on PTFE and from 32 to 125 mils on G-10.

The loss tangent of PTFE is assumed to be 0.0004 and the loss tangent of the G-10 epoxy-glass substrate is assumed to be 0.008. The effective Q due to the dielectric is equal to the inverse of the loss tangent. For PTFE this is 2,500 and it is evident that the majority of the loss is due to the conductors. Increased size reduces the conductor current density and results in increased unloaded Q. For epoxy-glass the dielectric unloaded Q is only 125. Increased resonator size results in little unloaded Q improvement.

Notice that distributed resonators require nearly 100 times the volume of a solenoid to achieve the same unloaded Q. However, distributed resonators are capable of providing higher Q_u. The highest practical solenoid Q is about 300. If size is no object, distributed resonators provide unloaded Qs of a thousand or more. Much of the volume of distributed resonators is due to length, particularly at lower frequencies.

The expressions for loss in coax are given in section 3.19 and the unloaded Q as a function of loss is given in section 3.15. The unloaded Q for coax in Figure 3-33 is based on these relationships. The expressions for loss in microstrip are more complex. The program =TLINE= has been used to compute the unloaded Q for the microstrip curves in Figure 3-33.

3.41 Discontinuities

The realization of filters using distributed elements necessarily involves the interconnection of physical structures. At higher frequencies, discontinuities such as line ends, steps in line widths, bends, and tee or cross junctions affect the behavior of these structures. The behavior of filters can be predicted by including lumped element discontinuity models in the overall structure.

Discontinuity models sufficiently accurate for practical filter development are known for popular processes such as microstrip, stripline and coaxial. Model accuracy is ultimately compromised with increasing operating frequency or physical size. Improved accuracy is obtained at higher frequencies and unusual geometries are simulated by electromagnetic modeling made feasible by economic digital computing. Electromagnetic simulation is orders of magnitude less computationally efficient than circuit simulation, so both methods are important.

Approximate circuit models for some commonly encountered discontinuities are shown in Figure 3-34. Note that these are approximate models; the models used in =SuperStar= and

=M/FILTER= described in Chapter 6 are generally more complex and may be frequency dependent. One important facet of any model is the set of "reference planes" used. These planes define what area is actually covered by the discontinuity model and may vary between model definitions. The planes used in =SuperStar= are shown with each discontinuity model in the reference chapter of the =SuperStar= manual.

Once the discontinuities have been identified, they must be corrected. This generally involves changing line lengths and may involve changing line impedances (especially for the tee and cross discontinuities which behave as if they had an embedded transformer). =M/FILTER= automatically compensates for, or absorbs, any discontinuities in the filters which it designs. The simplest discontinuity to absorb is the stray field open end effect on a stub. These stray fields are modeled by increasing the stub's length by a small amount (referred to here as L). Of

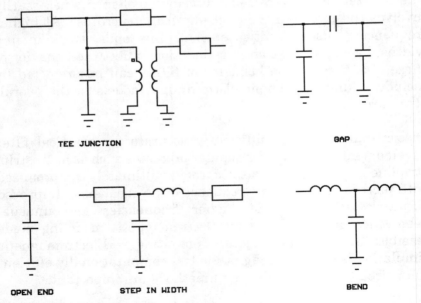

Figure 3-34 *Typical lumped-element models for planar circuit discontinuities. Some element values are frequency dependent in more accurate models.*

course, this does not mean that the line should actually be lengthened or that a capacitor should be added to the stub which the circuit is constructed; the lengthening just predicts how the original stub will perform. This means that an open-circuited stub will behave as if it were slightly longer than the actual physical stub and will resonate at a correspondingly lower frequency. To correct for this effect the stub must actually be shortened by ΔL to bring the behavior back to the designed behavior. This is illustrated in the following:

Original Designed Line Length: L
Length of line corresponding to end effect: ΔL
Apparent length including end effect: $L+\Delta L$
Length of line necessary including end absorption: $L-\Delta L$
Apparent corrected line length with end effect: $(L-\Delta L)+\Delta L=L$

Other discontinuities can be more tricky to absorb. For example, the width step correction not only absorbs extra line length but also absorbs the series inductor and shunt capacitor.

=M/FILTER= automatically compensates designs for all discontinuity effects. =M/FILTER= uses the full discontinuity models as simulated in =SuperStar= for maximum accuracy. The absorption process automatically takes place when the filter is calculated or when new electrical values are loaded in from an optimized electrical circuit or schematic file. =M/FILTER= lights the Absorb button in the flow diagram to inform the user that the compensation process is taking place. The absorption process corrects for optimal performance at the cutoff frequency for lowpass and highpass filters and at the center frequency for bandpass and bandstop filters. Since the discontinuities are frequency dependent, this may occasionally result in non-optimal performance at other frequencies. Optimization in =SuperStar= is used to correct for this effect.

3.42 References

[1] E.C. Jordon, ed., Reference Data for Radio Engineers: *Radio, Electronics, Computer, and Communications*, McMillan Publishing, Riverside, NJ, Seventh Edition, 1985, p. 6-13.

[2] S.P. Morgan, Effect of Surface Roughness on Eddy Current Losses at Microwave Frequencies, *J. Appl. Phys.*, Vol 20, 1949, p. 352.

[3] F.W. Grover, *Inductance Calculations*, Dover Publications, NY, 1962.

[4] Bureau of Standards Circular No. C74, Superintendent of Documents, Government Printing Office, Washington.

[5] H.A. Wheeler, Simple Inductance Formulas for Radio Coils, *Proc. IRE*, October 1928, p. 1398, and March 1929, p. 580.

[6] R.G. Medhurst, H.F. Resistance and Self-capacitance of Single-layer Solenoids, *Wireless Engineer*, February 1947, p. 35, and March 1947, p. 80.

[7] RCA Application Note #48, June 1935.

[8] E.C. Snelling, *Soft Ferrites: Properties and Applications*, Iliffe Books, London, 1969, p. 183.

[9] *Q Curves for Iron Powder Cores*, Micrometals, Anaheim, CA, undated.

[10] *RF/Microwave Capacitors Catalog and Data Book*, AVX Corporation, Myrtle Beach, SC, undated.

[11] S. B. Cohn, Beating a Problem to Death, *Microwave Journal*, November 1969, p. 22.

[13] N. Marcuvitz, *Waveguide Handbook*, Peter Peregrinus, LTD., London, 1986, p. 396.

[14] G. Stracca, G. Macchiarella, and M. Politi, Numerical Analysis of Various Configurations of Slab Lines, *Trans. MTT-34*, March 1986, p. 359.

[15] Substrates for Frequency Dependent Applications, Arlon Microwave Materials Division, 1100 Governor Lee Road, Bear, DE 19701, TEL (800) 635-9333.

[16] S.B. Cohn, Characteristic Impedance of the Shielded-Strip Transmission Line, *Trans. MTT-2*, July 1954, p. 52

[17] W. Hilberg, From Approximations to Exact Relations for Characteristic Impedances, *Trans. MTT-17*, May 1969, p. 259.

[18] H. A. Wheeler, Transmission Line Properties of a Stripline Between Parallel Planes, *Trans. MTT-26*, November 1978, p. 866.

[19] H. Howe, Jr., *Stripline Circuit Design*, Artech House, Dedham, MA, 1974, p. .

[20] S.B. Cohn, Shielded Coupled-Strip Transmission Line, *Trans. MTT-3*, October 1955, p. 29.

[21] R.K. Hoffman, *Handbook of Microwave Integrated Circuits*, Artech House, Dedham, MA, 1987, p. 124.

[22] T.G. Bryant and J.A. Weiss, Parameters of Microstrip Transmission Lines and of Coupled Pairs of Microstrip Lines, *Trans. MTT-16*, December 1968, p. 1021.

[23] T.G. Bryant and J.A. Weiss, MSTRIP (Parameters of Microstrip) -- Computer Program Description, *Trans. MTT-19*, p. 418.

[24] S. Yamashita and M. Makimoto, Miniaturized Coaxial Resonator Partially Loaded with High-Dielectric-Constant Microwave Ceramics, *Trans. MTT-31*, September 1983, p. 697.

[25] G. Stracca and A. Panzeri, Unloaded Q-Factor of Stepped-Impedance Resonators, *Trans. MTT-34*, November 1986, p. 1214.

[26] A. Zverev, *Handbook of Filter Synthesis*, John Wiley, New York, 1967, p. 499.

[27] D. Kajfez and P. Guillon, eds., *Dielectric Resonators*, Vector Fields, Oxford, Mississippi, 1990, p. 3.

[28] G. Matthaei, L. Young, and E.M.T. Jones, *Microwave Filters, Impedance-Matching Networks, and Coupling Structures*, Artech House, Dedham, Massachusetts, 1980.

[29] H.M. Barlow and A.L. Cullen, Microwave Measurements, *Constable*, 1950, Sec. 8.2.3.

[30] W.A. Edson, Microwave Filters Using Ghost-Mode Resonance, *IRE Electronics Comp. Conf.*, 1961, vol. 19, p. 2.

[31] G. Craven, Waveguide below Cutoff: A New Type of Microwave Integrated Circuit, *Microwave Journal*, August 1970, p. 51.

[32] G.F. Craven and C.K. Mok, The Design of Evanescent Mode Waveguide Bandpass Filters for a Prescribed Insertion Loss Characteristic, *Trans. MTT-19*, March 1971, p. 295.

[33] R.V. Snyder, New Application of Evanescent Mode Waveguide to Filter Design, *Trans. MTT-25*, December 1977, p. 1013.
[34] J. Bybokas and B. Hammond, High Temperature Superconductors, *Microwave Journal*, February 1990, p. 127.

4

Transformations

As we have seen, designing L-C lowpass filters from the lowpass prototype involves only scaling of the resistance and cutoff frequency of the prototype by simple multiplication and division. The transfer characteristics of the prototype are exactly retained and no special realization difficulties are introduced.

The design of other structures, such as highpass, bandpass and bandstop filters, require transformation in addition to the scaling. These transformations naturally modify the attributes of the prototype and may introduce severe realization difficulties, especially for bandpass and bandstop structures. The "ideal" transformation does not exist, and it becomes necessary to consider alternative transformations and how they relate to specific filter requirements and applications. This chapter considers a number of network transformations and equivalences.

4.1 Highpass Transformation

When $1/s$ is substituted for s in the lowpass transfer function, a highpass response results. However, instead of synthesizing a highpass prototype, it is only necessary to convert lowpass prototype series inductors to shunt capacitors and prototype shunt capacitors to series inductors while inverting the prototype values.

$$C_{hp} = \frac{1}{L_{lp}} \tag{1}$$

$$L_{hp} = \frac{1}{C_{lp}} \tag{2}$$

These normalized highpass values are then scaled to the desired resistance and frequency using equations (65) and (66) in Chapter 2.

4.2 Bandpass Conventional Transformation

Bandpass filters are also designed by transformation and scaling the lowpass prototype. We will consider several alternative bandpass transformations. With the conventional bandpass transform, each lowpass prototype series inductor transforms into a series inductor and a series capacitor. Each prototype shunt capacitor transforms into a shunt inductor and a shunt capacitor. Therefore the bandpass has twice the number of reactive elements as the prototype. The bandpass transfer function has twice the order[1] of the parent prototype. Shunning rigor, we will refer to the bandpass order as the order of the parent prototype.

For the bandpass, the lower cutoff frequency, f_l, and the upper cutoff frequency, f_u, are specified. We then define the center frequency, f_o, the absolute bandwidth, BW, and fractional bandwidth, bw, as

---

[1]*The order of a filter is the degree of the numerator of the voltage attenuation coefficient polynomial, H. The conventional bandpass transformation doubles the degree of the polynomial and therefore from a rigorous viewpoint, it doubles the order of the filter. However, the number of resonators and branches equals the number of branches in the lowpass prototype.*

$$BW = f_u - f_l \tag{3}$$

$$bw = \frac{BW}{f_o} \tag{4}$$

The percentage bandwidth is the fractional bandwidth multiplied by 100%. The transformed shunt element values are

$$C_{bpshunt} = \frac{C_{lp}}{bw} \tag{5}$$

$$L_{bpshunt} = \frac{1}{C_{bpshunt}} \tag{6}$$

and the transformed series element values are

$$C_{bpseries} = \frac{1}{L_{bpseries}} \tag{7}$$

$$L_{bpseries} = \frac{L_{lp}}{bw} \tag{8}$$

where C_{lp} is a lowpass prototype capacitor g-value and L_{lp} is a lowpass prototype inductor g-value.

The normalized bandpass values are scaled to the desired resistance and frequency using the denormalizing equations (65) and (66) in Chapter 2 with f_o as the frequency variable. The bandpass filter structures which result from the transformation of a 3rd order lowpass prototype are given in Figure 4-2. Figure 4-2a is the structure which results from the transformation of the lowpass prototype with a shunt capacitor at the input. For odd order, the number of inductors in the prototype is one less than the number of capacitors. The bandpass which results from the transformation of the lowpass prototype with a series inductor first is depicted in Figure 4-2b.

The transmission, reflection and group-delay responses of a 6th order, 0.177 dB passband ripple Chebyshev bandpass with a center frequency of 100 MHz and a bandwidth of 15% are given in Figure 4-1. The transmission response is plotted on the left grid with the far left scale of –100 to 0 dB. The reflection response (return loss) is also plotted on the left grid using the scale on the right of –30 to 0 dB. The group-delay is plotted on the right grid. These responses were computed and the plots were created using a digital computer and software described in

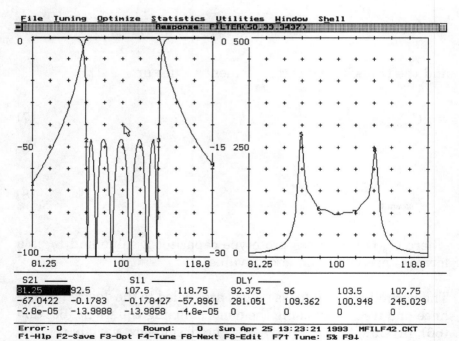

Figure 4-1 The transmission, reflection (return loss) and group-delay responses of a conventional bandpass filter with a bandwidth of 15% and centered at 100 MHz.

a later chapter. For these filter plots, the components are assumed lossless and parasitic-free.

Recall that the group-delay peaked at the corner of the lowpass response. With the bandpass, peaking occurs at both the lower and upper corners. The absolute value of the group-delay is inversely proportional to the absolute bandwidth and is independent of the center frequency. Notice that the amplitude transmission response exhibits greater selectivity on the low side of the passband and the resulting greater delay at the lower corner.

The bandpass transform involves simple mathematics, results in "exact" component values, and has an easily calculated amplitude response. Despite these desirable attributes, examination of equations 4.6 through 4.9 reveals that as the fractional bandwidth is decreased, the ratio of shunt branch and series branch inductor values (and capacitor values) becomes extreme[1]. For equal prototype values, a 10% bandwidth results in inductor and capacitor ratios of 100:1, resulting in severe realization difficulty. For narrow bandwidth, the conventional bandpass transform becomes impractical. Given in Table 4-1 are the component values of a 3rd order Butterworth 100 MHz bandpass terminated in 50 ohms with 40, 20, 10 and 5% bandwidth. The values of L_3 and C_3 in the 3rd shunt branch equal the values of L_1 and C_1. Notice the ratio of the shunt and series elements as the bandwidth is decreased. At 5% bandwidth, the ratio is 800:1.

It is difficult to overemphasize how significantly bandpass filter properties are related to bandwidth. The first mental calculation experienced filter designers make when presented with a filter specification is the percentage bandwidth.

---

[1]The impedance level of the bandpass may be adjusted to reduce the large series inductors or increase the small shunt inductors. However, the ratios are unaffected and decreasing the series inductors also decreases the already small shunt inductors. A more desirable solution is discussed later.

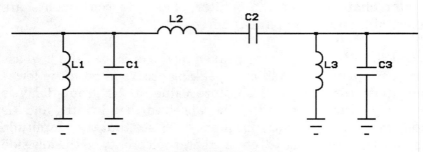

(a) MINIMUM INDUCTOR BANDPASS

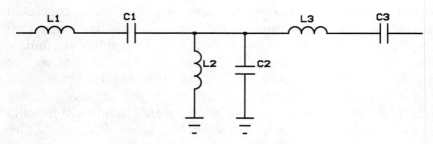

(b) MINIMUM CAPACITOR BANDPASS

Figure 4-2 *Bandpass filters created by conventional transformation of lowpass prototypes with (a) a shunt input capacitor and (b) a series input inductor.*

Component ratios, insertion loss, symmetry and component sensitivities are strongly impacted by the percentage bandwidth. The group delay is completely independent of the center frequency, but is inversely proportional to absolute bandwidth.

4.3 Bandstop Filter Transform

The bandstop filter has a series inductor and capacitor in shunt to ground for each shunt capacitor in the lowpass prototype and a parallel inductor-capacitor in series for each series inductor in the prototype.

The transformed shunt element values are

$$L_{bsshunt} = \frac{1}{C_{lp}bw} \tag{9}$$

$$C_{bsseries} = \frac{1}{L_{lp}bw} \tag{10}$$

and the transformed series element values are

$$C_{bsshunt} = \frac{1}{L_{bsshunt}} \tag{11}$$

$$L_{bsseries} = \frac{1}{C_{bpseries}} \tag{12}$$

The normalized bandstop values are scaled to the desired resistance and frequency using the denormalizing equations (65) and (66) in Chapter 2 with f_o as the frequency variable. The resulting bandstop filters are given in Figure 4-3 and the responses are given in Figure 4-4.

Two passbands exist with the bandstop. The lower passband extends from marker 1 to marker 2 on the left grid, and the

Table 4-1 *Conventional 3rd order Butterworth bandpass filter component values vs. percentage bandwidth.*

BW%	L1(nH)	C1(pF)	L2(nH)	C2(pF)	L3(nH)	C3(pF)
40	33.2	79.6	398	6.63	33.2	79.6
20	16.1	159	796	3.22	16.1	159
10	7.98	318	1592	1.59	7.98	318
5	3.98	637	3183	0.796	3.98	637

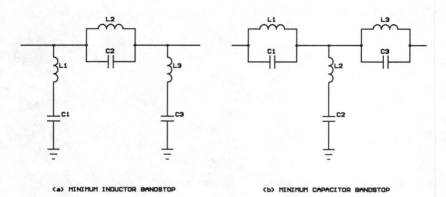

(a) MINIMUM INDUCTOR BANDSTOP   (b) MINIMUM CAPACITOR BANDSTOP

Figure 4-3 *Bandstop filters created by transforming a lowpass prototype with (a) a shunt input capacitor and (b) a series input inductor.*

upper passband extends from marker 3 to marker 4. The passband ripple of 0.177 dB manifests itself as return loss ripple of approximately 14 dB.

4.4 Narrowband Bandpass Transforms

The conventional bandpass transform results in a structure with alternating series and shunt resonators. Alternating resonator types are difficult to realize with certain classes of resonators. A number of other bandpass transformations have been developed to overcome this difficulty and the unrealizable component values of the conventional transform for narrow bandwidth.

Alternative transforms exist with all series or all parallel resonators which are based on impedance or admittance inverters. A quarter-wavelength line acts as an impedance or admittance inverter. L-C networks can also serve as inverters and may possess inversion properties over a wider bandwidth. These "like-resonator" bandpass filters are designed by creating a lowpass prototype with only series inductors. The shunt capacitors are converted to series inductors using impedance

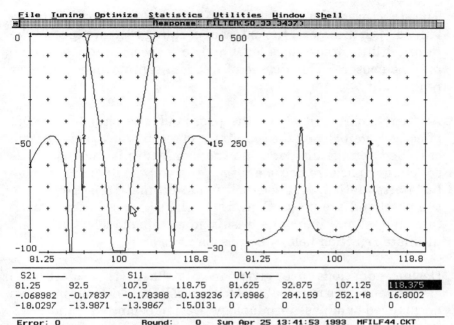

Figure 4-4 *The responses of a 6th order Chebyshev bandstop filter with 0.1777 dB passband ripple.*

inverters. Alternatively a prototype with all shunt capacitors may be created using admittance inverters. Impedance and admittance inverter theory is discussed in Section 4-20.

These design procedures yield only approximate component values and the response is found with increased computational difficulty. However, the "like resonators" and more realizable component values make these structures desirable for narrow bandwidth applications. An excellent reference with additional information on these filters is Matthaei [1].

4.5 Top-C Coupled, Parallel Resonators

One such structure consists of parallel L-C resonators in shunt to the transmission path. The resonators are coupled internally and externally by series capacitors at the top of the resonators.

The structure is depicted in Figure 4-5a. All inductors are equal valued. Furthermore, the external coupling reactors control the internal filter impedance level and so a specific inductor value may be chosen. This degree of freedom significantly enhances realizability.

At frequencies well above the passband, the shunt inductors effectively vanish and the coupling and resonator capacitors form cascaded voltage dividers. As the bandwidth is increased, the coupling capacitors become larger, the voltage dividers provide little attenuation, and selectivity and ultimate rejection above the passband are poor. Conversely, the series coupling capacitors and shunt inductors result in excellent selectivity and ultimate rejection below the passband.

Design equations are given in Matthaei [1]. Matthaei's expressions were manipulated to be consistent with our previous normalized bandpass terminology. Design begins with the selection of the normalized inductance of the inductors. For example, for a filter terminated in 50 ohms, choosing 100 ohm reactance inductors yields $L = 2$. Then the total normalized node capacitance at each resonator, C_t, is

$$C_t = \frac{1}{L} \tag{13}$$

The normalized coupling capacitors are then

$$C_{n,n+1} = \frac{bwC_t}{\sqrt{g_n g_{n+1}}} \tag{14}$$

where n indexes the shunt resonators sequentially and ranges from 1 to N. The input coupling capacitor is

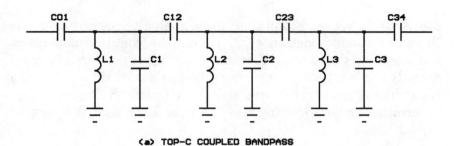

(a) TOP-C COUPLED BANDPASS

(b) TOP-L COUPLED BANDPASSS

Figure 4-5 *Top-C coupled and top-L coupled bandpass filters created using admittance inverter theory.*

$$C_{01} = \frac{J_{01}}{\sqrt{1-(R_a J_{01})^2}}$$

(15)

where

$$J_{01} = \left(\frac{C_t bw}{g_0 g_1}\right)^{0.5}$$

(16)

$C_{N,N+1}$ is found using the same expressions with appropriate indexes. R_a and R_b are the normalized input and output terminating resistance, respectively, typically 1.

The actual capacitance placed in parallel with each inductor is the total node capacitance, C_t, minus the coupling capacitors connected to that node. Internal coupling capacitors are used directly while external coupling capacitors and their terminating resistances are first converted to parallel R-C models to determine the node loading capacitances. Therefore, the parallel capacitors are

$$C_N = C_t - \frac{C_{N,N+1}}{C_{N,N+1}^2 + 1} - C_{N-1,N} \tag{17}$$

$$C_1 = C_t - \frac{C_{01}}{C_{01}^2 + 1} - C_{12} \tag{18}$$

$$C_n = C_t - C_{n-1,n} - C_{n,n+1} \tag{19}$$

and likewise for $C_{N,N+1}$. The normalized resistance and reactive values are then scaled to the desired resistance and frequency as before.

The design expressions are approximate and errors increase with increasing bandwidth. For this reason, and because of response asymmetry, this transform becomes unsuitable above about 20% bandwidth. Therefore, this and other similar structures are referred to as approximate narrowband filters. Above about 5% bandwidth, correction factors due to Cohn [2] and reviewed by Matthaei [1] are highly recommended.

The response of a 6th order, 0.177 dB Chebyshev, 15% bandwidth top-C coupled bandpass filter centered at 100 MHz is given in Figure 4-6. Greater selectivity and group delay occur below the passband. In this case, the attenuation at the lowest sweep frequency is 75.5 dB while the attenuation at the highest sweep frequency is only 47.8 dB. Likewise, the differential group-delay on the low side is twice the differential delay on the high side. The asymmetry is worse than the conventional bandpass and worsens with increasing bandwidth.

4.6   Top-L Coupled, Parallel Resonator

When greater attenuation is required above the passband, the structure given in Figure 4-5b may be used. It is similar to the previous structure except coupling inductors are used instead of coupled capacitors. The response is given in Figure 4-7. Design methods are similar to the top-C coupled filter and are covered in additional detail in Matthaei [1].

Notice that the asymmetry in the top-L coupled bandpass is not as extreme as with the top-C coupled filter. In fact, for 2nd

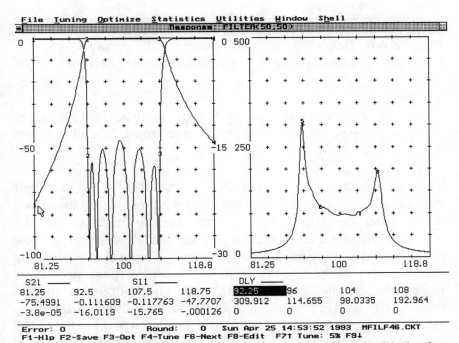

Figure 4-7 *Responses of a 6th order 0.177 dB ripple Chebyshev top-C coupled bandpass with 15% bandwidth centered at 100 MHz.*

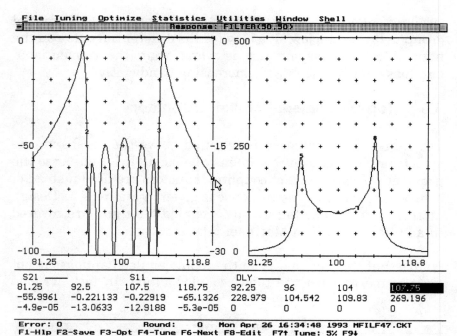

Figure 4-6 *Responses of a 6th order 0.1777 ripple Chebyshev top-L coupled bandpass filter of 15% bandwidth centered at 100 MHz.*

order, the symmetry in the top-*L* coupled filter is quite good. The asymmetry of both the top-*C* and top-*L* coupled filters worsens with increasing order.

The top-*C* and top-*L* coupled structures are excellent narrowband filters and provide alternatives when increased attenuation on either the lowside or the highside of the passband is desired. A disadvantage of these approximate narrowband structures is poor economy. Notice there are three elements for each reactive element in the lowpass prototype, plus an additional element. The two end coupling elements vanish with an appropriate parallel inductor value, but this may not be a desirable value from a realization viewpoint. One of the advantages of the top-*C* coupled structure is that the number of inductors is minimized.

As the bandwidth is decreased, the impedance of the series coupling reactors in the top-C and top-L coupled filters becomes quite high. While this poses no special problem for small capacitors in the top-C, for the top-L the series inductors become large and realization is troublesome. The unloaded Q of the series reactors is not as critical as the unloaded Q of the reactors in the parallel resonators.

4.7   Shunt-C Coupled, Series Resonator

An approximate narrowband structure designed with impedance inverters which results in series resonators is given in Figure 4-8a. This form uses shunt coupling capacitors to ground. Alternatively, shunt coupling inductors may be employed. The response of the shunt-C coupled series resonator bandpass is identical to the top-L coupled parallel resonator given in Figure 4-7. It provides increased attenuation above the passband while minimizing the number of inductors. Design procedures are again similar to the top-C coupled filter and are given in Matthaei [1].

The series resonators are typically less desirable from a construction viewpoint because both ends of the inductor are above ground potential. Also, parasitic capacitance to ground at the node between the series inductor and capacitor can be problematic.

4.8   Tubular Structure

The tubular filter is derived from the shunt-C coupled series resonator bandpass by first splitting the resonator series capacitors into two capacitors, one on each side of the inductors. The resulting "tee" of capacitors is then converted to the exact equivalent "pi". Because the tee and pi are exact equivalences, the response of the original shunt-C coupled series resonator bandpass and the tubular structure are identical. The schematic of a 3rd order tubular bandpass is given in Figure 4-8b.

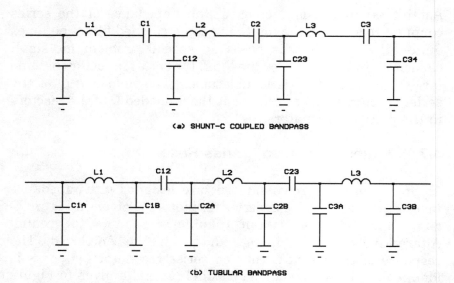

(a) SHUNT-C COUPLED BANDPASS

(b) TUBULAR BANDPASS

Figure 4-8 *Shunt-C coupled bandpass created using impedance inventors and the tubular structure created from the shunt-C using exact tee to pi network equivalences.*

$N-1$ additional capacitors are required but the floating nodes between series inductors and capacitors of the shunt-C coupled filter are eliminated. Therefore, every node has capacitance to ground which can be reduced to absorb stray capacitance at these nodes. Like the shunt-C coupled bandpass, all inductors are equal and the value is selectable within broad limits. The input and output shunt capacitors modify the effective termination resistance thus allowing control of the internal impedance and the inductor value. The ratios of these external capacitors are adjusted if necessary to provide for dissimilar terminations.

This structure is the basis of commercially popular tubular bandpass filters where it is realized in an axial mechanical form with shunt capacitors formed from short, low-impedance, coaxial element slugs, as depicted in Figure 4-9. The series capacitors are formed by a dielectric spacer between the flat faces of the

/ **WIRE**

▨ **DIELECTRIC**

■ **CONDUCTOR**

Figure 4-9 *Tubular bandpass filter structure with coaxial slugs for shunt and series capacitance, and axially wound helical inductors.*

slugs. The series inductors are helically wound on a dielectric support rod along the axial length of the filter and connected to adjacent slugs.

This structure has several desirable performance and realization attributes. Notice the inductors are naturally shielded from each other by the slugs. The capacitors are leadless and parasitic free. The "housing" is narrow which results in a high cut-off frequency and excellent stopband performance. The filter is tuned by compressing or spreading the inductors through a hole in the side of the housing. These holes are covered with a label or another tube.

4.9 Elliptic Bandpass Transforms

The previous bandpass transforms were applied to all-pole transfer function approximations. In the elliptic bandpass transformation, each finite transmission zero in the elliptic lowpass prototype results in two finite transmission zeros geometrically centered below and above the bandpass passband.

4.10   Conventional Elliptic Bandpass

The physical structure resulting from the conventional elliptic bandpass transformation has a parallel L-C in shunt for each lowpass prototype shunt capacitor and two parallel L-C networks cascaded in series for each parallel L-C in the series branches of the elliptic lowpass prototype, and is given in Figure 4-10. There is one inductor for each lowpass prototype branch plus an additional inductor for each lowpass prototype finite transmission zero.

Much like the conventional all-pole bandpass transformation, this structure has extreme ratios of element values with decreasing bandwidth. Also, the response is threatened by parasitic capacitance to ground at the floating node between the L-C pairs in the series branch. Design equations for the elliptic bandpass transformation are given by Williams [3].

Given in Figure 4-11 are the responses of a 6th order Cauer-Chebyshev elliptic bandpass filter with 0.177 dB passband ripple, 50 dB A_{min}, and 15% bandwidth centered at 100 MHz. Notice that the transition regions are much steeper than the previous 6th order all-pole responses. The attenuation limit of 50 dB is achieved much closer to the passband edges. The increased selectivity has resulted in increased differential group-delay. When comparing the selectivity of all-pole and elliptic filters, it should be recognized that for a given order the elliptic filter has additional components. If all-pole and elliptic filters with an equal number of elements are compared, the superiority of the elliptic must be judged on a case by case basis.

4.11   Zig-Zag Transformation

Saal and Ulbrich [4] presented design equations for an improved elliptic bandpass transform which saves an inductor for each lowpass prototype finite transmission zero. The transform is only available for even order lowpass prototypes.

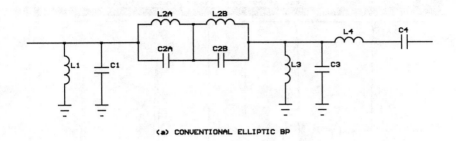

(a) CONVENTIONAL ELLIPTIC BP

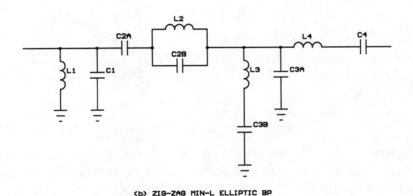

(b) ZIG-ZAG MIN-L ELLIPTIC BP

Figure 4-10 *Conventional and minimum-inductor zig-zag elliptic bandpass filters of 4th order.*

No bandpass structure provides more selectivity per inductor than the zig-zag. Removing an inductor from series branches also reduces the parasitic capacitance problem. Tuning is easier than the conventional elliptic transform because when constructed with precision capacitors, tuning is accomplished by adjusting the internal inductors for the correct zero frequencies and finally adjusting the two end inductors.

A reason the zig-zag is not more prevalent is that finding component values by manual calculation is not worthwhile. Williams [3] gives equations suitable for 4th order, but at higher order, Saal and Ulbrich [4] and a computer are required. With

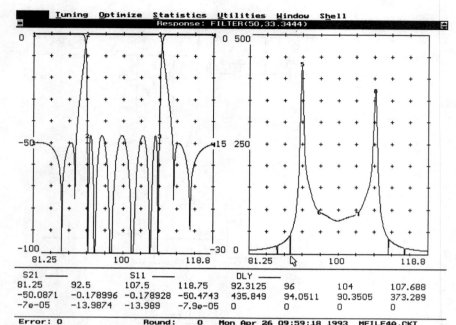

Figure 4-11 *Responses of a 6th order 0.177 dB Cauer-Chebyshev elliptic bandpass filter with a bandwidth of 15% centered at 100 MHz with a minimum stopband attenuation is 50 dB.*

economic computing today, this structure should become more popular.

4.12 Bandpass Transform Distortion

Arithmetic symmetry is often desired and results in group delay symmetry. Unfortunately, the conventional bandpass transform results in geometric amplitude symmetry and asymmetric group delay. None of the popular bandpass transforms result in arithmetic symmetry. The degree of asymmetry worsens with increasing bandwidth.

All published bandpass transforms also distort the phase and group delay attributes of the lowpass transfer function. For example, a bandpass designed by transforming a Bessel lowpass does not possess flat delay in the passband.

4.13 Arithmetic Transform

We have considered symmetry in the conventional, top-C coupled and top-L coupled bandpass transforms. The top-C coupled filter gives greater attenuation below the passband and the top-L filter gives greater attenuation above the passband. What structure results in arithmetic symmetry? Carassa [5] proved that arithmetic symmetry in a bandpass is possible if the number of transmission zeros in a structure at infinite frequency is three times the number at dc (multiplicity ratio is 3:1).

The conventional bandpass, the conventional elliptic bandpass and the zig-zag bandpass have an equal number of zeros at infinite frequency and dc. Therefore, the selectivity and group delay are greater below the passband. The multiplicity ratio for the top-C coupled filter is 1:$(2N+1)$ which is far less than 3:1 and the structure has very poor symmetry. The multiplicity ratio is $(2N+1)$:1 for the top-L coupled filter. This structure has a multiplicity ratio of 3:1 for the trivial case of $N = 1$ but the ratio increases with increasing order and therefore has greater selectivity and group delay above the passband.

Rhea [6] proposed a structure which has a multiplicity ratio of 3:1 for even order and asymptotically approaches 3:1 with increasing odd order. The topology of the symmetric transform is derived by using impedance inverters to transform the odd number shunt resonators of a conventional bandpass into series resonators. Shunt coupling capacitors are used as inverters. The even numbered shunt resonators and the series resonators in the conventional bandpass are not transformed. A bandpass resulting from the transformation of a 4th order lowpass prototype using this technique is given in Figure 4-12a. The amplitude and delay response of a 6th order symmetric transform bandpass is given in Figure 4-13.

The symmetry of this transform is excellent up to bandwidths as wide as 70%. Compare the transmission and group-delay symmetry in Figure 4-13 to the symmetry of the conventional and approximate narrowband filters shown in earlier figures.

Realizability issues are similar to the conventional transform. For bandwidths below 15% the ratio of maximum to minimum component values become extreme. This is avoided by using top-L coupled parallel resonator, shunt-C coupled series resonator or tubular bandpass filters for 15% and narrower bandwidths. These structures have reasonable symmetry at narrow bandwidths.

For bandwidths greater than 40%, impedance inverter theory is stressed and the symmetric bandpass develops unequal

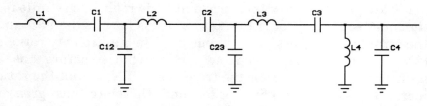

(a) SYMMETRIC TRANSFORM BP

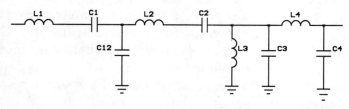

(b) BLINCHIKOFF FLAT DELAY BP

Figure 4-12 *4th order bandpass filters (a) created using the symmetric transform and (b) created by direct optimization of the bandpass transfer function for flat group-delay (due to Blinchikoff).*

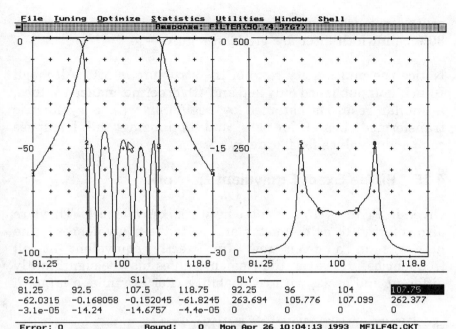

Figure 4-13 *Responses of a 6th order, 0.177 dB ripple, 15% bandwidth Chebyshev bandpass centered at 100 MHz created using the symmetric transform.*

passband ripple. This is recoverable to about 70% bandwidth by tuning the resonator frequencies.

4.14 Blinchikoff Flat Delay Bandpass

All published lowpass to bandpass transforms, including the symmetric transform just discussed, destroy the phase attributes of the lowpass. As with symmetry, the destruction worsens with increasing bandwidth.

Recognizing the transform as the culprit, Blinchikoff [7] synthesized a transfer function with nearly flat delay directly as a bandpass structure. He then published normalized values for 2nd and 4th order filters with 30 to 70% bandwidth. The 4th order structure is given in Figure 4-12b and the responses of a 30% bandwidth filter are given in Figure 4-14.

Notice the multiplicity ratio of the structure is 3:1. Although Blinchikoff published only 2nd and 4th order normalized values, a similar result is obtained by beginning with a symmetric transformed Bessel lowpass and optimizing the bandpass component values for flat delay.

4.15　Pi/Tee Exact Equivalent Networks

Given in Figure 4-15 are pi/tee network equivalences. They are also referred to as delta/star and delta/tee equivalences. The networks in a given row are exactly equivalent at all frequencies. However, the element values may be significantly different and assist with realizability. For example, a large C_c

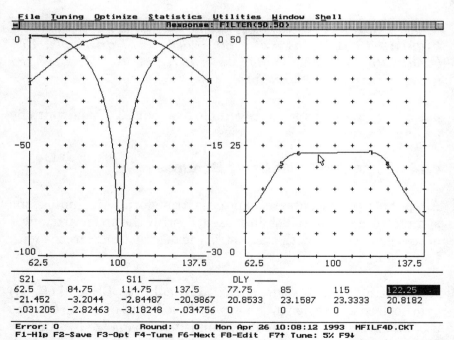

Figure 4-14 Responses of a 4th order Blinchikoff flat-delay 30% bandwidth bandpass filter centered at 100 MHz.

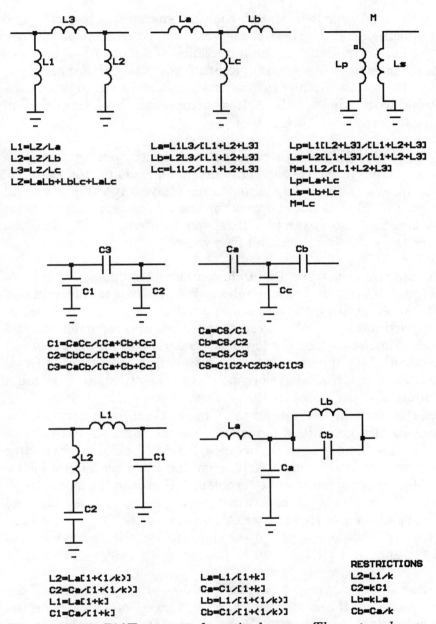

$L1=LZ/La$
$L2=LZ/Lb$
$L3=LZ/Lc$
$LZ=LaLb+LbLc+LaLc$

$La=L1L3/[L1+L2+L3]$
$Lb=L2L3/[L1+L2+L3]$
$Lc=L1L2/[L1+L2+L3]$

$Lp=L1[L2+L3]/[L1+L2+L3]$
$Ls=L2[L1+L3]/[L1+L2+L3]$
$M=L1L2/[L1+L2+L3]$
$Lp=La+Lc$
$Ls=Lb+Lc$
$M=Lc$

$C1=CaCc/[Ca+Cb+Cc]$
$C2=CbCc/[Ca+Cb+Cc]$
$C3=CaCb/[Ca+Cb+Cc]$

$Ca=CS/C1$
$Cb=CS/C2$
$Cc=CS/C3$
$CS=C1C2+C2C3+C1C3$

$L2=La[1+(1/k)]$
$C2=Ca/[1+(1/k)]$
$L1=La[1+k]$
$C1=Ca/[1+k]$

$La=L1/[1+k]$
$Ca=C1/[1+k]$
$Lb=L1/[1+(1/k)]$
$Cb=C1/[1+(1/k)]$

RESTRICTIONS
$L2=L1/k$
$C2=kC1$
$Lb=kLa$
$Cb=Ca/k$

Figure 4-15 *Pi/Tee network equivalences. The networks on a given row are exactly equivalent. The component values in the bottom row networks are not independent.*

in the capacitor tee network could be susceptible to series lead inductance. The largest capacitor in the equivalent pi network is smaller, minimizing the inductance problem. An equivalent form may have other manufacturing advantages. For example, the tee to pi equivalence was used in Section 4.8 to create the tubular bandpass which has exceptional performance and construction attributes.

All element values in a given network for the purely capacitive and purely inductive networks are independent. For example, C_1, C_2 and C_3 in the capacitive pi may have any value not equal to zero. This is not the case for the mixed capacitor/inductor networks. For example, three parameters, L_1, C_1 and the arbitrary factor k define all four values in the network.

These relations were used to find all four bandpass networks in Figure 4-16 which are equivalent. When driven and terminated in 50 ohms, they are narrowband bandpass filters centered at approximately 1 GHz. On the upper left, the capacitors form a tee. The network on the lower left is derived from the upper left network by applying the tee to pi transform given in Figure 4-15. Although the network responses are identical, the maximum value of capacitance in the pi form is less than a fifth of the capacitance in the tee form. This substantially reduces the effects of a given lead inductance for the capacitors. With 0.5 nH of lead and ground inductance, the 53.33 pF shunt capacitor series resonates at 975 MHz, near the filter center frequency, causing severe realizability problems! However, the shunt 10 pF capacitors in the pi equivalent network would resonate well above the center frequency at 2.25 GHz. If the 10 pF capacitors are reduced to 8.4 pF to compensate for the inductive lead reactance at 1 GHz, the filter passband is nearly recovered.

The equivalence may be applied to 100% or only a portion of the elements on the extremities of a network. The network on the upper right in Figure 4-16 is created by first splitting the original series 16 pF capacitors in the network on the upper left into two series capacitors, one of 40 pF and one of 23.33 pF. The

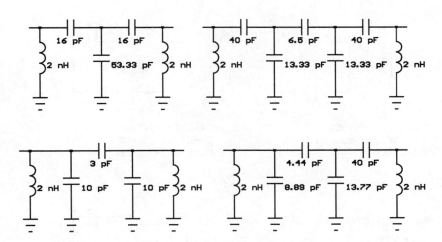

Figure 4-16 *Tee and pi network equivalent examples.*

resulting tee with 23.33 pF series capacitors and the 53.33 pF shunt capacitor is then converted to a capacitor pi with 13.33 pF shunt capacitors and a 6.5 pF series capacitor. The resulting network has two additional capacitors (it is non-canonic), but additional control over element values is available. In the network on the lower right, all of the original 16 pF series capacitor on the left is transformed while only a portion of the original 16 pF series capacitor on the right was transformed. The resulting network has one additional capacitor.

4.16 Exact Dipole Equivalent Network

Given in Figure 4-17 are important equivalent dipoles (two-terminal networks used as branches in a more complex structure). For the top dipoles, the form on the left has a smaller inductor and larger capacitors. The dipole on the top right and repeated on the lower left has the form of the equivalent electrical model of many piezoelectric resonators such as bulk-quartz crystals, two-terminal SAW resonators and ceramic resonators. The dipole on the lower right has an additional capacitor, C_c, with an arbitrary user selectable value. The inclusion of this capacitor provides for L_a smaller than L_1

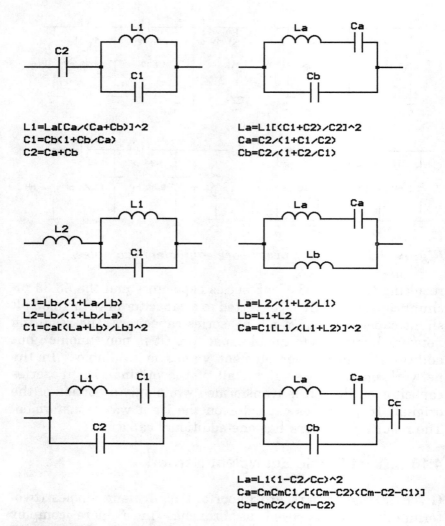

L1=La[Ca/(Ca+Cb)]^2
C1=Cb(1+Cb/Ca)
C2=Ca+Cb

La=L1[(C1+C2)/C2]^2
Ca=C2/(1+C1/C2)
Cb=C2/(1+C2/C1)

L1=Lb/(1+La/Lb)
L2=Lb/(1+Lb/La)
C1=Ca[(La+Lb)/Lb]^2

La=L2/(1+L2/L1)
Lb=L1+L2
Ca=C1[L1/(L1+L2)]^2

La=L1(1-C2/Cc)^2
Ca=CmCmC1/[(Cm-C2)(Cm-C2-C1)]
Cb=CmC2/(Cm-C2)

Figure 4-17 *Exact equivalent network dipoles. The top two dipoles are canonic. The bottom dipoles require an additional capacitor but offer control of the inductance value.*

and with a controllable value. Zverev [8] includes a listing of all possible two-inductor, two-capacitor, dipole equivalent networks.

This is also an excellent reference for other equivalent networks and transformations.

4.17 Norton Transforms

Norton developed one of the most useful equivalent transform classes. Their usefulness is due to the ability to control the impedance level in a network by adding a reactor while avoiding a two-winding transformer. Norton's first transform operates on a series reactor and his second transform operates on a shunt reactor. Two forms of Norton's first transform are given in the top two rows of Figure 4-18 and two forms of his second transform are given in the bottom two rows. The K factor in Figure 4-18 is the turns ratio of the transformer. K may be greater or smaller than 1. The transformer impedance ratio is K^2. The resistor symbol represents reactance. This reactor is either a capacitor or an inductor. If the original series reactor is an inductor, the equivalent circuit is given by the schematic in the center of the top row. If the original series reactor is a capacitor, the equivalent circuit is given by the schematic on the top right.

The use of Norton's transformations are best illustrated by an example. Recall the 20% bandwidth, 100 MHz center frequency Butterworth conventional bandpass filter terminated in 50 ohms given in Figure 4-2a. The ratio of shunt to series inductor values is almost 50 to 1. This filter is repeated in Figure 4-19a.

In Figure 4-19b, ideal step-down and step-up transformers are placed before and after the series resonator to lower the transmission system impedance for the series resonator. $K=7.035$ transforms the impedance down by 49.498. This reduces the original 796 nH inductor to equal the shunt 16.1 nH inductors and increase the original series capacitor to 159 pF to equal the shunt capacitors. Next, to eliminate the difficult and expensive transformer, Norton's second transform is applied to the shunt 16.1 pF capacitor and the transformer. The rightmost capacitor in the resulting Norton tee is negative. To preserve

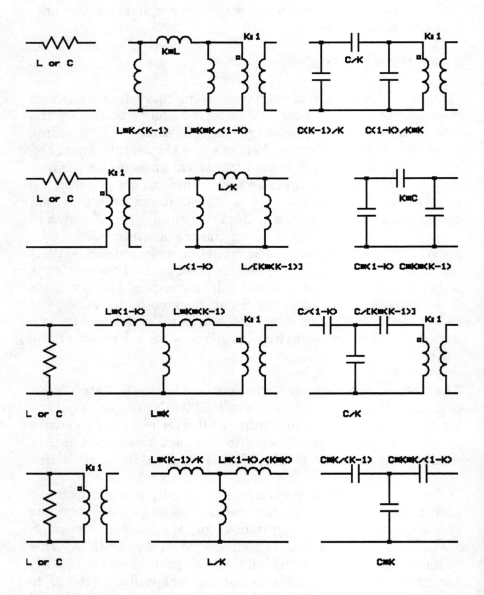

Figure 4-18 *Norton's first (top two rows) and second transformations (bottom two rows) replace a reactor and an impedance transformer with three reactors. One is negative and is absorbed into an adjacent element.*

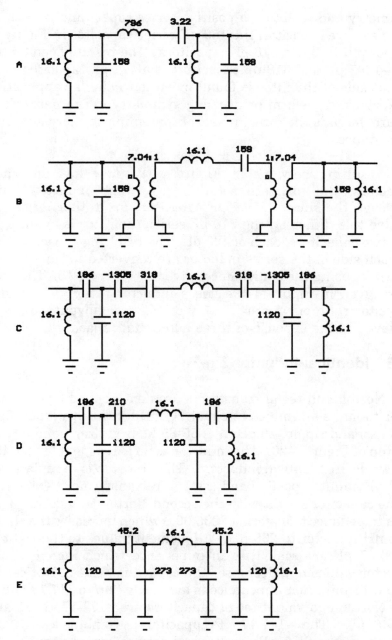

Figure 4-19 *Norton and tee to pi transformation of a conventional bandpass filter to reduce the inductor values ratio.*

symmetry and to obtain a positive series capacitance to absorb this negative capacitor, the 159 pF series capacitor in Figure 4-19b is split into two 318 pF capacitors. The result of combining the −1305 pF and 318 pF capacitors is a 421 pF capacitor. The output side of the filter is found by either a similar application of Norton's transform or utilizing symmetry. The network in Figure 4-19d is the direct result of combining the series branch capacitances.

The resulting network in Figure 4-19d now has all equal inductors, the ultimate solution to a high inductor ratio. However, the shunt 1119.7 pF capacitors are rather large. To resolve this difficulty, the 210 pF series capacitor is again split into two capacitors, each of 420 pF. The resulting capacitor tees on each side of the series inductor are converted to capacitor pi networks using the equivalences given in Figure 4-15. The final filter given in Figure 4-19e has a inductor ratio of 1:1 and a capacitor ratio of approximately 6:1. These advantages were achieved at the expense of three additional capacitors.

4.18 Identical-Inductor Zig-Zag

The Norton and tee/pi transforms are next applied to a zig-zag filter to make all inductors equal. Consider the 4th order, 0.11 dB passband ripple, 40 dB A_{min}, 55-85 MHz zig-zag bandpass at the top of Figure 4-20. The inductor ratio is reasonable at 4.4:1, but we desire identical inductors. The first step is to replace the 94.7 pF shunt capacitor and a step-up transformer (K=0.6517) to the capacitor's right with the second Norton transform. This is an impedance transform of 235/99.8 which increases the shunt 99.8 nH inductor to 235 nH. All element values to the right of the 94.7 pF are scaled up in impedance which increases the inductor values and decreases the capacitor values. Notice the output termination impedance is also scaled up to 117.17 ohms. The Norton equivalent tee of capacitors are −177.17, 61.71 and 115.46 pF. The −177.17 pF capacitor is combined with the original series 25.2 pF capacitor to form a series capacitor of 29.38 pF. This tee is then converted to a pi of capacitors as

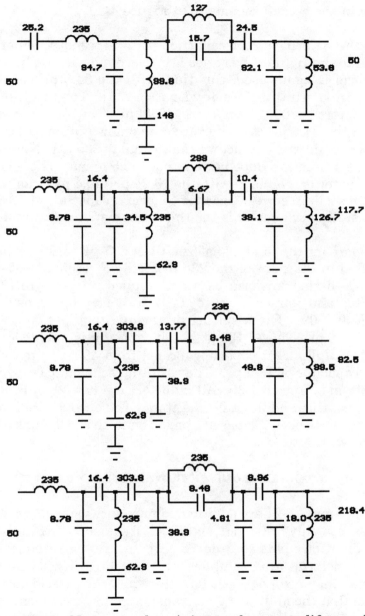

Figure 4-20 *Norton and tee/pi transforms modify an elliptic zig-zag bandpass to an exact equivalent bandpass with additional capacitors but with identical inductors.*

shown in the second schematic in Figure 4-20.

Next, the 34.5 pF shunt capacitor in the second schematic is shifted to the right of the series L-C to ground. Norton's second transform is again used, this time with the 34.5 pF capacitor and a step-down transformer with $K=1.1281$. The resulting tee of capacitors is 303.8, 38.9 and −342.74 pF. The elements to the right of the 34.5 pF capacitor are scaled down in impedance by 235/299 so the third inductor is now equal to 235 nH. Notice the output termination impedance is now 92.5 ohms. The −342.74 pF capacitor is combined with the 13.24 pF (10.4 pF prior to the 235/299 scaling) series capacitor to form a capacitor of 13.8 pF. The resulting schematic is the third schematic in Figure 4-20.

To prepare for the final transform, the 13.77 pF series capacitor is shifted to the right of the 235 nH/8.48 pF resonant network. Norton's first transform is then applied to this 13.77 pF capacitor and immediately to its right a step-up transformer with $K=0.6509$. The resulting pi of capacitors is 4.808, 8.964 and −3.130 pF. The −3.130 pF shunt capacitor is combined with the 21.085 pF shunt capacitor (scaled from 49.8 pF) to form a shunt capacitor of 18.0 pF. The resulting schematic is given at the bottom of Figure 4-20. All inductors are now 235 nH at the expense of three additional capacitors. The output termination resistance has been increased from 50 ohms to 218.4 ohms which is dealt with next.

4.19   Approximate Equivalent Networks

Given in Figure 4-21 are additional impedance transforms which are exact only at the specified frequency. Since the approximation holds to a degree over a range of frequencies, these transforms are useful for bandpass filter structures. The fact that they are not exact for all frequencies is evidenced by the fact that the multiplicity of transmission zeros is modified at either dc or infinite frequency.

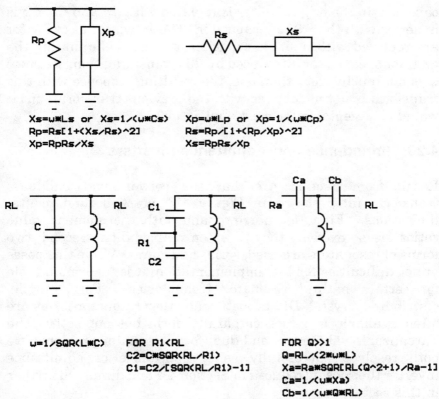

Xs=w*Ls or Xs=1/(w*Cs)
Rp=Rs[1+(Xs/Rs)^2]
Xp=RpRs/Xs

Xp=w*Lp or Xp=1/(w*Cp)
Rs=Rp/[1+(Rp/Xp)^2]
Xs=RpRs/Xp

w=1/SQR(L*C)

FOR R1<RL
C2=C*SQR(RL/R1)
C1=C2/[SQR(RL/R1)-1]

FOR Q>>1
Q=RL/(2*w*L)
Xa=Ra*SQR[RL(Q^2+1)/Ra-1]
Ca=1/(w*Xa)
Cb=1/(w*Q*RL)

Figure 4-21 *Approximate transformations useful over a limited bandwidth.*

When the capacitive tap transformation is applied to the bottom schematic in Figure 4-20 to transform the output impedance down to 50 ohms, C=18.1 pF, RL=218.4 ohms, and $R1$=50 ohms. The resulting $C1$=34.7 pF and $C2$=37.8 pF. The resulting filter now has four additional capacitors but all inductors are identical and both the input and output terminations equal the original 50 ohms.

The approximate transform on the lower right in Figure 4-21 also may be used to drop the 218.4 ohm output termination to 50 ohms. To satisfy the requirement for Q>1, Cb is on the filter side and Ca is on the termination side of L. Because the 18.1 pF capacitor in the original filter is replaced with Ca and Cb, Cb is in series with the 8.96 pF capacitor. These two series capacitor are replaced with a single capacitor, therefore eliminating the additional capacitor introduced by this transformation. Because Q is not much larger than one, the resulting response with this transform is not as ideal as with the previous transform, but it would be acceptable for most applications.

4.20 Impedance and Admittance Inverters

Recall from Section 4.2 that the conventional bandpass transform filter shown in Figure 4-2 has two fundamental limitations. First, for narrow bandwidth the element value ratios are so extreme they threaten realizability. Second, two forms of resonators are used; series and shunt. While this poses minor difficulties for L-C implementation, it is insurmountable for certain natural resonator technologies. For example, distributed, crystal, DR, acoustic and other resonator forms are often available in series or parallel form but not both. The unrealizable value ratio and dual form resonator problems are both resolved successfully using impedance or admittance inverter theories introduced in Section 4.4 and discussed further in this section.

Suppose a bandpass structure is desired using shunt parallel resonators only. If a number of parallel resonators are cascaded directly, there exists an equivalent network which is a single shunt resonator with capacitance equal to the sum of all resonator capacitances and an inductor equal to the parallel equivalence of all inductors. Resonators of alternating form are required to eliminate this problem, which is the idea struck upon by both Wagner and Campbell who discovered multiple section filters. An alternative solution is to cascade similar resonators but isolate them with impedance or admittance inverters.

One form of impedance inverter is a quarter wavelength transmission line. When terminated with an impedance Z_b, the input impedance of a quarter wavelength line is

$$Z_a = \frac{Z_o^2}{Z_b} \tag{20}$$

Besides inverting the impedance of the termination, inverters must have a phase shift of ±90 degrees, or a multiple thereof.

Likewise, admittance inverters are defined by the relationship

$$Y_a = \frac{Y_o^2}{Y_b} \tag{21}$$

A series inductor with an admittance inverter on each side looks like a shunt capacitor. Likewise, a shunt capacitor with an impedance inverter on each side looks like a series inductor. Therefore, the lowpass prototypes shown in Figure 2-6 may consist entirely of shunt capacitors or series inductors cascaded with inverters as shown in Figure 4-22. The reactors in Figure 4-22 may have arbitrary values. The impedances of the quarter wavelength inverters are

$$Z_{01} = \sqrt{\frac{R_a L_1}{g_o g_1}} \tag{22}$$

$$Z_{n,n+1} = \sqrt{\frac{L_n L_{n+1}}{g_n g_{n+1}}} \tag{23}$$

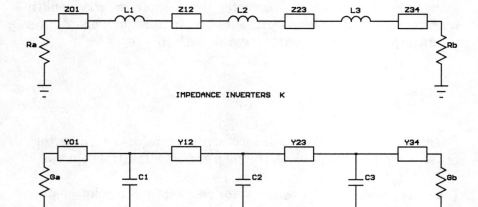

IMPEDANCE INVERTERS K

ADMITTANCE INVERTERS J

Figure 4-22 *Impedance (K) inverters convert shunt capacitors to series inductors. The prototype then consists of series inductors and inverters. Also shown is a converted prototype using admittance inverters.*

$$Z_{N,N+1} = \sqrt{\frac{R_b L_N}{g_N g_{N+1}}} \tag{24}$$

The admittances of the quarter wavelength inverters are given by

$$Y_{01} = \sqrt{\frac{G_a C_1}{g_0 g_1}} \tag{25}$$

$$Y_{n,n+1} = \sqrt{\frac{C_n C_{n+1}}{g_n g_{n+1}}} \tag{26}$$

$$Y_{N,N+1} = \sqrt{\frac{G_b C_N}{g_N g_{N+1}}} \tag{27}$$

Quarter wavelength transmission lines provide inversion over a limited bandwidth. Wider bandwidth inversion results when lumped element impedance and admittance inverters are used. A number of practical inverters are given in Figure 4-23. For lumped element inverters, the impedance inverter parameter K is substituted for Z_o in equation (20) and the admittance inverter parameter J is substituted for Y_o in equation (21). Negative lumped elements and transmission line lengths are realizable provided the inverters are placed such that those elements are absorbed into adjacent positive elements.

When normal bandpass transformations are applied to the structures in Figure 4-22, the results are bandpass filters with similar resonator form. For example, if the impedance inverter in the top right of Figure 4-23 is used with series inductors bandpass transformed into series L-C resonators, the structure given in Figure 4-8a is obtained. The limited bandwidth of the impedance and admittance inverters limits how faithfully the desired transfer approximately is reproduced as the desired filter bandwidth is increased. Filters designed using impedance or admittance inverter theory are best applied to narrowband filters. For L-C filters this is not overly unfortunate because the shortcomings of conventional bandpass filters vanish with increasing bandwidth. Inverters may be applied against resonators as well as against reactors as with the distributed elliptic bandpass filter discussed in Section 8-24.

4.21 Richards Transform

In Section 3-18 the approximate equivalence of a shorted transmission line and a lumped inductor was described. This equivalence is good at all frequencies where the electrical line length is much less than 90 degrees. A more rigorous concept

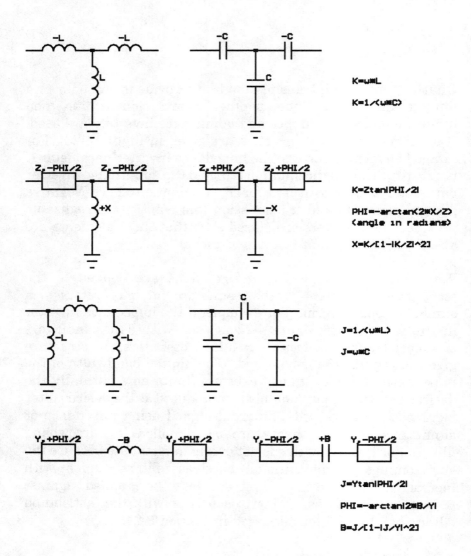

Figure 4-23 *Impedance and admittance inverter structures. X is the reactance, B the susceptance of reactors. Z is the impedance, Y the admittance and PHI the radian length of line elements.*

developed by Richards [9] provides a theoretical basis for applying the principles of exact L-C filter synthesis to distributed filters. All line lengths are 90 degrees long at f_o, the design frequency (not the cutoff frequency). The Richards transform then maps the frequency domain variable f to the Richards variable, Ω, in a new domain often referred to as the S-plane. The mapping is

$$S = j\Omega = j\tan\frac{\pi f}{2f_o} \tag{28}$$

This transformation is used by a number of contributors [10-16] to exactly synthesize distributed filter structures without resorting to approximate line/lumped equivalence.

4.22 Kuroda Identities

Listed in Figure 4-24 are Kuroda's transformations and derivatives thereof [13]. These transformations include a unit transmission line element (u.e.) with an electric length of 90 degrees at the design frequency and a characteristic impedance of Z or admittance Y. The transformations also include one or more lumped elements.

To illustrate the use of Richards and Kuroda transforms we will design a distributed 2.3 GHz 3rd order 0.10 dB ripple Chebyshev lowpass filter for a 50 ohm system. The stopband center occurs at f_o, the frequency at which the transmission line elements are 90 degrees long. The first reentrance passband is centered at $2f_o$. A reasonable choice for f_o is $2f_c$. A higher f_o results in wider stopbands but greater transmission line impedance ratios. We will chose f_o=5.5 GHz. Therefore

$$\Omega_c = \tan\frac{2.3\pi}{2\times5.5} = 0.771 \tag{29}$$

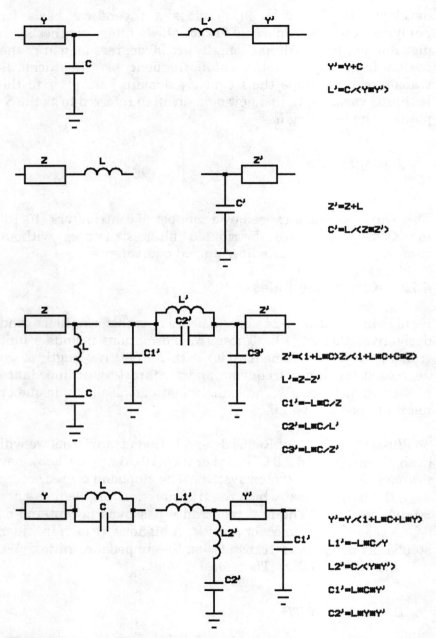

Figure 4-24 *Kuroda's and Kuroda-Levy transformations involving a unit element and one or more lumped reactors.*

The distributed lowpass is realized as shorted-series stubs for the lowpass prototype series inductors and open-shunt stubs for the prototype capacitors. The impedance of the series stubs and the admittance of the shunt stubs are given by

$$Z_i = \frac{g_i R}{\Omega_c} \tag{30}$$

$$Y_i = \frac{g_i R}{\Omega_c} \tag{31}$$

The lowpass prototype reactive g-values from the Table 2-5 are

$$g_1 = g_3 = 1.0316 \tag{32}$$

$$g_2 = 1.1474 \tag{33}$$

Using the lowpass prototype form beginning with a series inductor, the impedances of the two series stubs lines are

$$Z_1 = Z_3 = \frac{1.0316 \times 50}{0.771} = 66.89 \, ohms \tag{34}$$

and the admittance of the shunt stub is

$$Y_2 = \frac{1.1474 \times 0.02}{0.771} = 0.0298 \, mhos \tag{35}$$

The impedance of the shunt stub Y_2 is therefore 33.60 ohms.

This completes the design for manufacturing processes which allow series stubs such as wire-line. However, series stubs are unrealizable in microstrip. To overcome this difficulty, we use the Kuroda transform in the second row of Figure 4-24 to convert the series inductors into shunt capacitors. The unit element may not be initially inserted within the filter next to the series element because the internal impedance is not in general equal to Z_o. However, the unit elements may be added externally and shifted within the filter by repeated application of Kuroda transforms. The steps are illustrated in Figure 4-25.

The inductors in the top schematic represent series shorted transmission line elements with a characteristic impedance of 66.89 ohms and length of 90 degrees at 5500 MHz as determined above. The capacitors represent shunt open stubs of 0.0298 mhos (33.56 ohms) and length 90 degrees at 5500 MHz. In the second schematic, unit elements with $Z=50$ ohms and 90 degree length at 5500 MHz are added. They have no effect on the amplitude response and add only linear phase. In the third schematic, a Kuroda transform is used to convert the series stubs to shunt stubs. From Figure 4-24

$$Z' = Z + L = 50 + 66.89 = 116.89 \tag{36}$$

$$C' = \frac{L}{ZZ'} = \frac{66.89}{50 \times 116.89} = 0.0114 \tag{37}$$

The 116.89 ohm lines are no longer series stubs but are conventional transmission lines. In the fourth schematic in Figure 4-25, the shunt stubs are redrawn as shunt open stubs, which completes the design.

Richards and Kuroda transforms, in combination with sophisticated exact synthesis techniques, offer the promise of precise filter design for any requirement. However, a number of problems conspire to destroy this ideal view, such as discontinuities, limited line impedance ratios and reentrance. Also, with increasing order, the repeated application of Kuroda transforms to shift unit elements into the filter worsens the impedance ratio problem. As will be discussed in Chapter 6, although far less elegant, brute force numeric techniques using a digital computer offer hope in managing these issues.

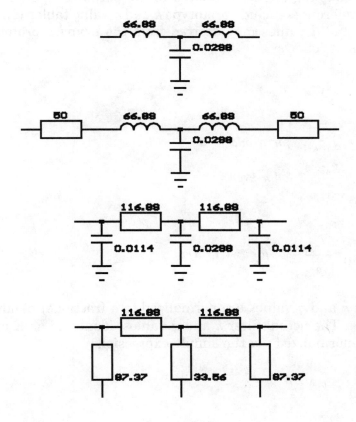

Figure 4-25 *A distributed lowpass filter developed from the lumped lowpass prototype using Richards and Kuroda transforms.*

4.23 Prototype k and q Values

Bandpass filters may be defined by only three entities; a resonator structure, coupling between resonators (internal coupling) and coupling to the terminations (external coupling). A straightforward design procedure based on these concepts begins with a prototype defined by k and q values. k_{ij} represents the coupling between resonators i and j, q_1 is related to the input coupling and q_n is related to the output coupling. Some filter references include prototype k and q value tables [8]. Also, the k and q values may derived from the lowpass prototype g values as follows

$$q_1 = g_0 g_1 \tag{38}$$

$$q_n = g_n g_{n+1} \quad for \ n \ odd$$
$$= \frac{g_n}{g_{n+1}} \quad for \ n \ even \tag{39}$$

$$k_{i,i+1} = \frac{1}{\sqrt{g_i g_{i+1}}} \quad for \ i=1 \ to \ n-1 \tag{40}$$

These k and q values are normalized to a fractional bandwidth of one. The actual filter k and q values, identified as K and Q, are denormalized via the simple expressions

$$K_{i,j} = k_{i,j} \frac{BW}{f_o} \tag{41}$$

$$Q_1 = q_1 \frac{f_o}{BW} \tag{42}$$

$$Q_n = q_n \frac{f_o}{BW} \tag{43}$$

where BW is the absolute bandwidth and f_o is the bandpass center frequency.

The denormalized K and Q values are then used with analytical expressions to design the bandpass filter by finding element values. Williams [3, p.5-19] uses this procedure to design top-C coupled bandpass filters of the form shown in Figure 4-5a. We will use K and Q values in Chapter 8 to illustrate a powerful technique for designing bandpass filters of almost arbitrary form.

4.24 References

[1] G. Matthaei, L. Young and E.M.T. Jones, *Microwave Filters, Impedance-Matching Networks, and Coupling Structures*, Artech House, Norwood, Massachusetts, 1980.

[2] S.B. Cohn, Direct-Coupled-Resonator Filters, *Proc. IRE*, Vol. 45, February 1957, p. 187.

[3] A.B. Williams and F.J. Taylor, *Electronic Filter Design Handbook*, McGraw-Hill, New York, 2nd ed., 1988.

[4] R. Saal and E. Ulbrich, On the Design of Filters by Synthesis, *IRE Trans. Circuit Theory*, Vol. CT-5, December 1958, p. 284.

[5] F. Carassa, Band-Pass Filters Having Quasi-Symmetrical Attenuation and Group-Delay Characteristics, *Alta Frequenza*, Vol. XXX, No. 7, July 1961, p. 488.

[6] R.W. Rhea, Symmetrical Filter Design, *RF Expo East 1989 Proceedings*, Cardiff Publishing, Englewood, Colorado, 1989.

[7] H.J. Blinchikoff and M. Savetman, Least-Squares Approximation to Wideband Constant Delay, *IEEE Trans. Circuit Theory*, Vol. CT-19, July 1972, p. 387.

[8] A.I. Zverev, *Handbook of Filter Synthesis*, John Wiley and Sons, New York, 1967, p. 522.

[9] P.I. Richards, Resistor-transmission-line circuits, *Proc. IRE*, vol. 36, February 1948, p. 217.

[10] R.J. Wenzel, Exact Design of TEM Microwave Networks Using Quarter-Wave Lines, *MTT-12*, January 1964, p. 94.

[11] M.C. Horton and R.J. Wenzel, General Theory and Design of Optimum Quarter-Wave TEM Filters, *MTT-13*, May 1965, p. 316.

[12] R.J. Wenzel, Exact Theory of Interdigital Band-Pass Filters and Related Coupled Structures, *MTT-13*, September 1965, p. 559.

[13] B.M. Schiffman and L. Young, Design Tables for an Elliptic-Function Band-Stop Filter (N=5), *MTT-14*, October 1966, p. 474.

[14] R. Levy and I. Whiteley, Synthesis of Distributed Elliptic-Function Filters from Lumped-Constant Prototypes, *MTT-14*, November 1966, p. 506.

[15] M.C. Horton and R.J. Wenzel, The Digital Elliptic Filter - A Compact Sharp-Cutoff Design for Wide Bandstop or Bandpass Requirements, *MTT-15*, May 1967, p. 307.

[16] R.J. Wenzel, Synthesis of Combline and Capacitively Loaded Interdigital Bandpass Filters of Arbitrary Bandwidth, *MTT-19*, August 1971, p. 678.

5

Filter Losses

The ideal filter transfers all incident energy at passband frequencies to the filter output termination. In practice, energy is lost by reflection at the filter ports, dissipation within the filter and/or radiation from the filter[1]. These topics are considered in this chapter.

5.1 Reflection or Mismatch Loss

Reactive filters provide transition region and stopband rejection by reflecting energy at the input and returning that energy to the source. Certain filter transfer functions, such as Chebyshev, reflect some energy even at frequencies well into the passband in order to achieve greater transition region steepness (selectivity).

Passband reflections may also occur unintentionally. For example a mismatched output termination may reflect energy which passes back through the filter and is lost in the source termination.

If only the source or load is mismatched, the mismatch loss, L_m, for a purely reactive network is

$$L_m = -10\log(1 - |\rho|^2) \tag{1}$$

---

[1]*We will use the term insertion loss to signify lost energy which does not arrive at the output termination for any of these reasons. Thus insertion loss may be desired (the stopband) or unintentional (dissipative losses).*

where ρ is the reflection coefficient. For example, the mismatch loss of an ideal lowpass filter terminated at the output in 100 ohms instead of a nominal 50 ohms is -10 times the base 10 logarithm of $1-.333^2$ which is 0.51 dB. The reflection coefficient is $(VSWR-1)/(VSWR+1)$ where the $VSWR$ is 100/50 or 2.0. Equation (1) may also be used to relate the passband reflection coefficient ripple in a Chebyshev filter to the attenuation ripple in decibels.

5.2 Unloaded Q Induced Loss

Practical lumped and distributed elements possess dissipative loss. Dissipation occurs in electric fields in lossy dielectrics and with current flowing in lossy conductors. Various component technologies have significantly different loss mechanisms and magnitudes, and certain circuit configurations are more susceptible to component losses. It is important to recognize the distinction between component and network induced loss effects.

Unloaded or component Q is defined in section 3.3 and the unloaded Q of specific reactor technologies is discussed throughout Chapter 3. Unloaded Q is the ratio of stored to dissipated energy in a reactor. It is related to material properties, and for a given material, to the physical size of the reactor. Energy dissipated in the reactor manifests itself as dissipative insertion loss in filter structures.

However, the same reactor may result in significantly different insertion losses in various filter structures. The reasons are discussed in the next sections.

5.3 Loaded Q Definitions

Loaded Q is a design parameter of bandpass and bandstop structures.

$$Q_l = \frac{f_o}{BW} = \frac{1}{bw_{frac}} \tag{2}$$

where f_o is the geometric center frequency, BW is the absolute bandwidth and bw_{frac} is the fractional bandwidth. The percentage bandwidth is 100% times bw_{frac}. The loaded Q is undefined for lowpass and highpass filters.

5.4 Lowpass Loss

Finite unloaded Q results in lowpass passband attenuation which generally increases as the corner frequency is approached. This results in response roll-off at the corners and has the effect of rounding the amplitude response.

At dc and low frequencies, the insertion loss in decibels due to finite unloaded Q for lowpass filters is

$$I.L._{dc} = \frac{4.34}{Q_{uave}} \sum_{n=1}^{n=N} g_n \tag{3}$$

where Σg_n is the sum of all reactive g-values in the prototype and Q_{uave} is the average inductor and capacitor unloaded Q. Notice the g-values, which are defined by the transfer approximation and order, and the unloaded Q of the components are the only factors which determine the dissipative insertion loss. The dissipative insertion loss is inversely proportional to the component unloaded Q.

Interestingly, the shape insertion loss as a function of frequency is nearly proportional to the shape of the group delay response if component Q is modeled as a constant series resistance for inductors and a constant parallel resistance for capacitors.

Given on the right half in Figure 5-1 are the transmission amplitude and group delay responses for a 7th order Chebyshev lowpass filter with unloaded component Qs of 500. The loss is modeled as constant resistance in series with the inductors and as constant resistance in parallel with the capacitors.

The response on the right is the dissipative insertion loss caused by finite-Q components. The dissipative insertion loss curve was generated by computer simulation by subtracting S_{21} for the lossy filter from S_{21} for an identical but lossless lowpass filter. The scale of the group delay and dissipative insertion loss plots were adjusted to simplify comparison. Notice the amazing similarity. The dissipation loss peaks very near the frequency of maximum group delay and asymptotically approaches a finite value at low frequencies, as does the group delay. In the transition region, the loss decreases at a rate faster than the group-delay. The group delay characteristics of the lowpass filter provides a simple and clear visualization of the nature of finite-Q dissipation loss. The low-frequency value of dissipation loss as predicted by equation (3) can be multiplied by the relative group-delay to estimate the dissipative loss versus frequency. The relationship between the group-delay and dissipative loss exists for bandpass filters as well.

5.5   Bandpass Loss

For a single resonator, the insertion loss at resonance due to finite unloaded Q is

$$I.L._o = 20\log\frac{Q_u}{Q_u - Q_l} \tag{4}$$

For multi-section bandpass filters, the band center insertion loss in decibels due to finite unloaded Q is

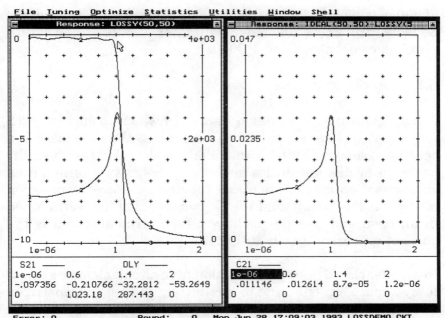

Figure 5-1 *On the left are transmission magnitude and group delay responses for a 7th-order Chebyshev lowpass with finite-Q components and on the right is the dissipative loss.*

$$I.L._o = \frac{4.34Q_l}{Q_u} \sum_{n=1}^{n=N} g_n \tag{5}$$

Again, the shape of the insertion loss as a function of frequency is proportional to the bandpass group delay response.

Notice an important distinction. For the lowpass, once a prototype is selected, only the unloaded Q and the selected prototype affect the loss. If Q_u is large the loss is small. However, with the bandpass, dissipation loss is a function of bandwidth. Even if Q_u is high, narrow bandwidth filters have significant insertion loss. For example, Σg for a 5th order Butterworth is 6.472. $I.L._o$ with an unloaded to loaded Q ratio of 300 is 0.094 dB, for a ratio of 100 is 0.28 dB and for a ratio of 30 is .94 dB. For a 10% bandwidth filter, $I.L._o$ less than 0.1 dB requires component Q near 3000.

5.6 Radiation Loss

A homogeneous transmission line, even when unshielded such as microstrip, has very little radiation unless the strip to ground spacing is a significant fraction of a wavelength. However, the presence of discontinuities such as open ends or termination shorts can induce radiation and substrate surface propagation modes. It has been found that a transmission line terminated in an open end is a particularly effective radiator and is the basis of microstrip patch antennas.

In an unenclosed network, radiation from circuit elements is lost to free space, contributing to a reduction in energy transferred from the network input to the network output (loss). Parasitic coupling between elements due to radiation is minimal in enclosed housings as long as the operating frequency is below the cutoff frequency of waveguide modes within the housing. At frequencies above cutoff, propagation modes within the housing cause coupling between radiating elements[1]. This parasitic coupling can significantly perturb the circuit behavior, causing either an increase or decrease in signal levels as viewed from the network terminals depending on the vector summation of signals from desired and parasitic paths.

To improve the accuracy of circuit simulation, over the last few decades considerable effort has been invested in analytical models and electromagnetic simulation of distributed elements. It is a frequent topic of discussion during evaluations of the merits of various simulator programs. The author's experience is that for many filter configurations the impact of radiation,

---

[1] *Radiation and reception are reciprocal. An element which radiates energy into the enclosure can receive energy from the enclosure.*

which is not considered in any circuit simulator, may overwhelm the significance of other discontinuity effects.

While by no means comprehensive, examples of the impact of radiation on circuit performance are investigated in this section. Measured responses under different conditions are used to justify hypothetical explanations involving radiation. A rigorous, integrated, approach to the simulation of radiation effects on circuit performance is currently beyond the scope of interactive simulation. However, the following discussion provides insight into typical problems and potential solutions.

5.7 Radiation from Microstrip Resonators

Easter and Roberts [1], expanding on previous referenced works, developed an expression for the radiated power fraction for half-wavelength open-ended microstrip resonators. The radiated power fraction is the ratio of radiated power, P_r, to the power incident, P_i, at each end of the resonator. Their expression (MKS units) is:

$$\frac{P_r}{P_i} = \frac{64\eta\,\pi h^2}{3\epsilon_{eff}\,Z_o\lambda_o^2} \tag{6}$$

where h is the substrate thickness and η is the intrinsic impedance of free space. For example, the power radiated fraction for a half-wavelength resonator formed with a 50 ohm line on 31 mil thick, 2.55 dielectric-constant board ($\epsilon_{eff} \approx 2.14$) at 5.6 GHz is 5.1%.

Circuit designers may not be accustomed to dealing with the concept of radiation. A more familiar term might be Q_r, the Q resulting from power lost to radiation. The power radiated fraction and Q_r are related by

$$Q_r = \frac{2\pi P_i}{P_r} \qquad (7)$$

Note that the power radiated fraction is inverted in this expression. For the above microstrip resonator case, Q_r is 123.2. The unloaded Q for the same resonator limited by copper and dielectric loss [2] with 0.71 mils metalization thickness, a loss tangent of 0.0004 and an RMS surface roughness of 0.06 mils is approximately 370. Radiation is clearly the predominant factor in limiting the unloaded Q of this resonator.

These considerations significantly modify conventional circuit design precepts. For example, when radiation is considered, the resonator line impedance[1] which results in maximum unloaded Q is no longer $77/\varepsilon^{1/2}$. Gopinath [3] considered total microstrip half-wave resonator unloaded Q including radiation. For small substrate thickness, radiation is insignificant and conventional conductor and dielectric loss considerations determine the unloaded Q. As the substrate thickness is increased, radiation becomes more prevalent and decreases the unloaded Q. Because radiation decreases with higher line impedance, for thicker substrates, the optimum line impedance for maximum unloaded Q tends to increase. Gopinath gives several nomographs relating these parameters.

5.8   Surface Waves

In addition to launching a radiated wave perpendicular to the strip conductor plane, when the operating frequency is above the housing cutoff, discontinuities excite surface waves (pre-

---

[1]*This relation holds only for coaxial resonators. For microstrip, the optimum line impedance is a function of additional parameters such as the metalization thickness. In general, lower impedances give optimum unloaded Q in microstrip and stripline.*

dominantly *TM* mode) which propagate over the substrate. Multiple surface waves launched from strip discontinuities experience reflection at all boundaries, resulting in complex parasitic coupling modes. The resulting ultimate rejection floor is stated by Hoffman [4] to be 60, 40 and 20 dB at 1, 5 and 10 GHz, respectively. An excellent review is given by Hoffman of the work by many contributors on radiation and package modes.

The above radiation and surface wave effects are considered in an isolated environment. Inclusion of these effects in an integrated, comprehensive, program is currently beyond the scope of circuit simulation for several reasons. Transmission line conductor and dielectric loss is definable per unit length while radiation is related to specific objects and lengths, such as a half-wave resonator. The radiation characteristics of non-resonant lines differ from resonant lines. Either analytical models for the radiation characteristics of all microstrip objects and discontinuities must be developed or electromagnetic radiation simulation, which is not fast enough to be interactive, must be employed. Even if analytical radiation models were available, three dimensional vector simulation or Fourier transforms in the circuit plane are required to determine radiation from the composite circuit. The circuit may not be considered in isolation because the package has significant impact on radiation effects as we will see in the next section.

5.9 Edge-Coupled Bandpass Radiation Examples

Given in Figure 5-2 is an =M/FILTER= synthesis program design screen for a 7th order microstrip edge-coupled Chebyshev bandpass filter centered at 5.6 GHz with 0.0432 dB passband ripple. The *PWB* has a nominal dielectric constant of 2.55, a thickness of 31 mils and a metalization thickness of 1.42 mils, a surface roughness of 0.06 mils *RMS* and a loss tangent of 0.001. =M/FILTER= is covered in Chapter 6 and the edge-coupled bandpass filter class is discussed in more detail in Chapter 8. The length of coupling between resonators is not the conventional 50% of the resonator length (90 degrees), but is slid

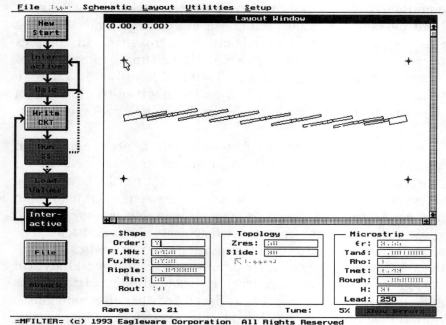

Figure 5-2 =M/FILTER= *design screen for a 7th-order edge-coupled bandpass filter.*

apart by a factor of 20 degrees. The layout angle is adjusted so that the input and output 50 ohm leader lines are on a horizontal axis.

The filter was constructed by etching the pattern on a 2.54 inch square *PWB*. The *PWB* was mounted in a 2.54 inch square aluminum housing which was 0.56 inches deep. A cover was not installed on the housing. The transmission amplitude and return loss responses were measured on a vector network analyzer and recorded on a flatbed plotter. The results are given in Figure 5-3.

A number of interesting observations are evident. Notice that the ultimate rejection well into the stopbands is only 45 to 50 dB, depending on the chosen reference; either 0 dB or the minimum passband attenuation point. The ultimate rejection is

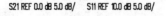

S21 REF 0.0 dB 5.0 dB/   S11 REF 10.0 dB 5.0 dB/

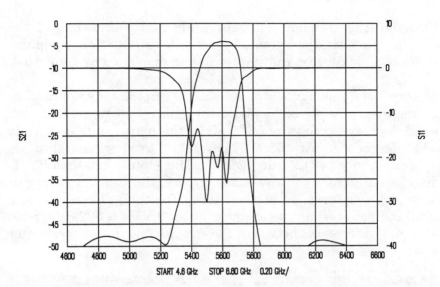

START 4.8 GHz   STOP 6.80 GHz   0.20 GHz/

Figure 5-3 *Transmission amplitude and return loss responses of a microstrip edge-coupled 5.6 GHz bandpass filter mounted in a 2.54 inch square housing.*

in general support of Hoffman's suggested 40 dB value at 5 GHz. Even more interesting is the passband rolloff on the low side. Notice both the return loss and the amplitude transmission just below 5400 MHz. At the low frequency cutoff where the return loss passes through 15 dB, the amplitude transmission loss is over 20 dB. The 15 dB return loss indicates very little energy is reflected but the transmission loss is substantial. The loss is far greater than would be expected from conductor and dielectric loss, as is evidenced by the transmission loss higher in the passband. What is causing this energy loss? A probable answer is radiation. However, the magnitude of radiation from a half-wave resonator predicted by equation (6) is not sufficiently large to account for the observed loss. Also, why is the loss so much larger on the low side of the passband? In fact, the mid-band

and high-side loss are what would be expected from conductor and dielectric loss. Why does radiation not occur midband?

Consider the transmission amplitude and phase-shift of the edge-coupled bandpass found by computer simulation (computer simulation is covered in Chapter 6) and displayed in Figure 5-4.

At very low frequency, the phase shift asymptotically approaches 90 degrees. Below 5225 MHz, the phase has rotated though −180 degrees, so that by 5225 MHz, the phase shift is already 360 degrees. At 5380 MHz the total phase-shift is approximately 481.2 degrees. The total phase-shift is distributed across the seven resonators, the input and output coupling structures, and the leaders. The total phase-shift in a similar single-resonator bandpass at 5380 MHz is 322.8 degrees. The net difference between the single and seven resonator

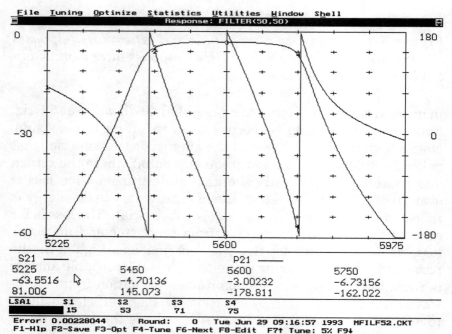

Figure 5-4 Transmission amplitude and phase-shift responses of the microstrip 5.6 GHz edge-coupled bandpass.

structure is therefore 158.4 degrees or 26.4 degrees per added resonator. A list of transmission phase-shift versus frequency for both the seven- and single-section filters is given in Table 5-1. Let's hypothesize that the edge-coupled bandpass is an array of resonator elements, each of which contributes to the total radiated power. From Table 5-1 we conclude that at 5300 MHz the relative phase-shift from resonator to resonator is relatively low and the radiated power in the far-field is nearly the in-phase sum of the individual components. However, at 5880 MHz the currents in adjacent radiating resonators are nearly 180 degrees out of phase with each other and the radiated energy cancels. This hypothesis is consistent with the unusual loss pattern on the low side of the edge-coupled passband but which vanishes as the frequency is increased. The radiation was also manifested as significant response variation with movement of a cover plate even when it was removed from the housing by several inches.

Table 5-1 *Transmission phase-shift versus frequency for the seven-section edge-coupled bandpass filter, a single-section edge-coupled and the phase-shift per resonator.*

FREQ	7-SECTION $S21\ \theta$	1-SECTION $S21\ \theta$	/SECTION
5300	404.3	310.2	15.7
5340	434.0	316.0	19.7
5380	481.2	322.8	26.4
5420	570.0	331.2	39.8
5460	693.5	341.7	58.6
5500	789.8	355.0	72.5
5540	872.7	371.7	83.5
5580	950.8	390.9	93.3
5600	988.8	400.6	98.0
5640	1065.9	418.7	107.9
5680	1147.6	432.0	119.3
5720	1241.3	445.3	132.7
5760	1335.1	454.6	146.8
5800	1447.7	462.0	164.3
5840	1484.3	468.3	169.3
5880	1506.7	473.7	172.2

To further test the hypothesis, the width of the edge-coupled filter *PWB* was trimmed and the *PWB* was mounted in a 0.75 inch wide housing. The cutoff frequency of the lowest waveguide mode of a 0.75 inch channel is 7869 MHz. Energy below this frequency propagates with severe attenuation in the housing which suppresses radiation from the *PWB*. The measured responses of the same edge-coupled bandpass mounted in the narrow housing are given in Figure 5-5.

When the filter is mounted in the 0.75 inch wide housing the amplitude response loss is moderate over the entire frequency range for which the return loss is 10 dB or better. The 4 dB mid-band insertion loss is in reasonable agreement with the computer-predicted value of 3 dB in Figure 5-4. The measured filter bandwidth is narrower than the computer predicted value.

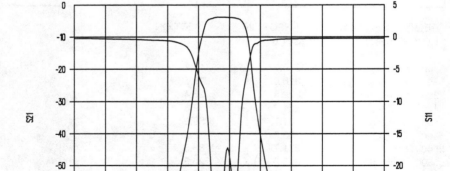

Figure 5-5 *Measured responses of the edge-coupled bandpass installed in a housing with a width of 0.75 inches.*

The narrower bandwidth would result in increased passband insertion loss.

Further evidence that mounting the filter in the 0.75 inch wide housing suppressed free-space radiation was elimination of response variation due to cover plate movement. Also, the narrow channel has suppressed surface-wave radiation as evidenced by the 15-20 dB increase in ultimate stopband rejection.

5.10 Hairpin Bandpass Radiation

Shown in Figure 5-6 is an =M/FILTER= synthesis program screen for a hairpin bandpass with the same electrical parameters as the unfolded edge-coupled in Figure 5-2. Folding

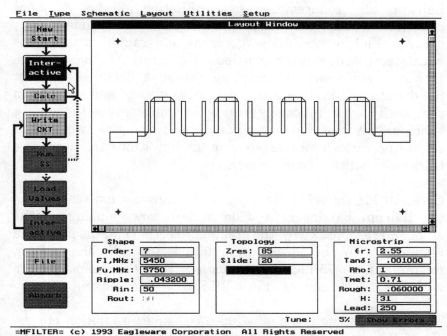

Figure 5-6 *Hairpin bandpass filter created by folding the straight edge-coupled structure and compensating the design for the bends and shortened coupling length.*

the resonators creates a more compact filter with a more rectangular aspect ratio. The total resonator length is still 180 degrees but the bends have electrical length and tend to shorten the physical length of the resonators. The longitudinal length of line is included in the total resonator length so that the remaining line lengths which couple to adjacent resonators are somewhat shorter. This decreases the spacing between resonators. Otherwise, edge-coupled and hairpin synthesis are identical. The edge-coupled and hairpin bandpass are discussed further in Chapter 8.

A dual form of the straight edge-coupled bandpass with open-circuit ends is an edge-coupled structure with shorted resonators. While this form is useful and has significantly lower radiation, the required via holes are generally considered a nuisance. Folding the edge-coupled open-circuit resonators into hairpins also reduces free-space radiation [5] due to phase cancellation of fields at the ends where radiation predominantly occurs. Field cancellation increases and radiation decreases with decreased spacing between the arms of the hairpin. However, self-resonator coupling causes a decrease in filter bandwidth and increases the loss. Also, design and simulation complexity are significantly reduced by using wide self-resonator spacings. A reasonable self-resonator spacing is two to four times the inter-resonator spacing, or five times the substrate thickness, whichever is greater.

Given in Figure 5-7 is the amplitude transmission response of the hairpin bandpass in wide and narrow housings. The stopband responses with the greatest rejection and the passband response with the lowest attenuation are with the hairpin mounted in a 0.75 inch wide housing. The other responses are with the hairpin mounted in a 2.54 inch wide housing.

The hairpin structure has reduced free-space radiation effects. The differences in the passband insertion loss responses when mounted in the narrow and wide housings are far less than with the straight edge-coupled bandpass. Even when mounted in a

S21 REF 0.0 dB 10.0 dB/

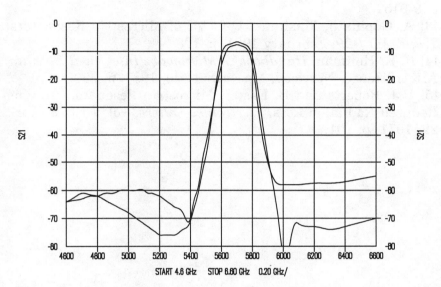
START 4.6 GHz   STOP 6.60 GHz   0.20 GHz/

Figure 5-7 *Transmission amplitude response of a 5.6 GHz hairpin bandpass mounted in a wide (plot with poorest stopband rejection and greatest insertion loss) and narrow housing.*

wide housing, the hairpin shows no evidence of the severe low side rolloff present in the edge-coupled bandpass. There is about 1 dB of additional insertion loss when the hairpin is mounted in a wide housing. Also notice the slight bandwidth difference. There is some evidence that folding the resonators into a hairpin does not eliminate surface-wave radiation because the narrow housing significantly increases the stopband rejection. However, even in the wide housing, the hairpin provides greater stopband rejection.

5.11   References

[1] B. Easter and R. Roberts, Radiation from Half-Wavelength Open-Circuit Microstrip Resonators, *Electronics Letters,* Vol. 6, No. 18, September 3, 1970, p. 574.

[2] =T/LINE= program, Eagleware Corp., 1750 Mountain Glen, Stone Mtn., GA, 30087, USA, TEL (404) 939-0156, FAX (404) 939-0157.
[3] A. Gopinath, Maximum Q-factor of Microstrip Resonators, *Trans. MTT-29*, February 1981, p. 128.
[4] R. K. Hoffman, *Handbook of Microwave Integrated Circuits*, Artech House, Norwood, Massachusetts, 1987, p. 390.
[5] R.J. Roberts and B. Easter, Microstrip Resonators Having Reduced Radiation Loss, *Electronics Letters*, Vol. 7 No. 8, April 22, 1971, p. 191.

6

Computer-Aided Strategies

More than easing computational burdens, the digital computer has revolutionized the way we design filters. Any modern treatment of filter design must address strategies which have become practical including real-time tuning, statistical analysis, sensitivity analysis, design centering and optimization. It is now feasible to optimize for desired and customized characteristics while simultaneously considering component losses, parasitics and discontinuities.

Many filter synthesis theories which we use today were developed in an age when computing tools were far less sophisticated[1]. Wonderfully elegant mathematical solutions were found for a variety of filter problems, but idealized assumptions were required to make the process manageable. Today, these idealized symbolic theories form a starting point which is followed by brute force numeric techniques.

6.1 Overview

In this chapter we will discuss digital computer techniques and software programs available for the development of filters. They may be classified in two broad categories; synthesis and simulation. Synthesis is the process of designing filters by finding the topology and component values or dimensions.

---

[1]*In the early 1960s, slide rules provided fewer than one floating point operations per second (Flops) with a precision of three significant digits. By the 1970s, the scientific calculator provided increased speed, precision and a larger function set. By 1981, 8 bit personal computers augmented with mathematics coprocessors provided 50 Kflops. By the early 1990s, inexpensive desktop computers achieved over 2 Mflops.*

Simulation is a process of evaluating a design by calculation and display of the filter responses.

On the upper left in Figure 6-1 are the transmission and return loss responses on an ideal L-C 7th order Chebyshev lowpass filter with a cutoff frequency of 2250 MHz and passband ripple of 0.177 dB. The transmission is plotted on a scale of 6 dB per division and the return loss is plotted on a scale of 3 dB per division. The schematic of the L-C filter is given in Figure 6-2a.

On the upper right in Figure 6-1 are the responses of an equivalent ideal transmission line stepped-impedance lowpass filter created by converting the lumped elements to distributed elements using the equivalences given in Figure 3-15. The shunt capacitors are converted to low impedance lines and the series inductors are converted to high impedance lines. The responses on the upper right of Figure 6-1 were computed using ideal transmission line elements without losses or discontinuities.

Lossy lines and the discontinuities associated with the steps in width cannot be avoided. The top view of a microstrip implementation is given in Figure 6-2b and a schematic representation of the microstrip filter is given in Figure 6-2c. The responses for this physical realization including losses and discontinuities are given on the lower left in Figure 6-1.

Both the lumped to distributed conversion and the physical implementation perturbed the ideal L-C responses. At 2250 MHz, the effects of the discontinuities are minimal but clearly evident. As the frequency or substrate thickness is increased, the effects of the discontinuities are increased. On the lower right in Figure 6-1 are the responses after optimization of the line lengths to achieve performance similar to the original L-C lowpass. Notice that the passband ripple and corner frequency have been recovered. However, the ultimate rejection at 4500 MHz is somewhat degraded.

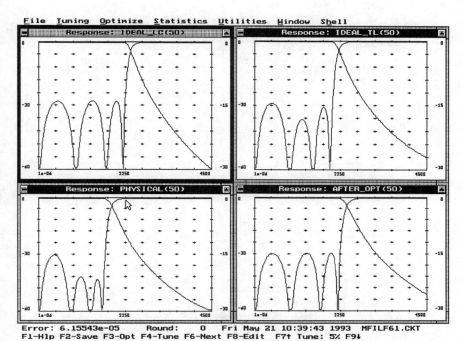

Figure 6-1 Transmission and return loss responses of a lowpass filter in ideal L-C, ideal transmission line, physical line and optimized physical line, respectively, left to right.

Next we will discuss the effects illustrated in Figure 6-1 for the stepped-impedance lowpass in more detail to illustrate computer-aided design techniques. The design of stepped-impedance and other distributed filters is discussed further in later chapters.

6.2 Synthesis CAE

The simple L-C lowpass filter in Figure 6-2a was created using the synthesis program =FILTER= [1]. Synthesized L-C structures in =FILTER= include all the transforms discussed in Chapter 4 and others listed in Table 6-1. All-pole and elliptic lowpass, highpass, bandpass and bandstop filters are synthesized based on lowpass prototype scaling and transformation techniques discussed in Chapters 2 and 4.

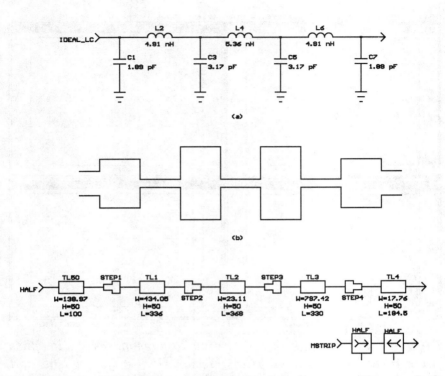

Figure 6-2 *2250 MHz cutoff Chebyshev lowpass filter in L-C form, (a), in microstrip form viewed from above, (b), and in a transmission line schematic form (c).*

Distributed structures can be designed using transmission line equivalences for the inductors and capacitors in L-C filters. However, for certain structures, synthesis procedures which recognize the special nature of distributed elements result in more effective and realizable designs. Also, distributed filter structures which require series capacitors should be avoided. The program =M/FILTER= includes algorithms and routines specifically appropriate to synthesize the microwave distributed filter structures discussed in detail in later chapters and listed in Table 6-2. Both =FILTER= and =M/FILTER= design filters with the transfer function approximations listed in Table 6-3.

After synthesizing a filter and displaying the schematic or layout on-screen, these programs automatically write simulator circuit files to describe the filter to be analyzed. =FILTER= writes a lumped element file and =M/FILTER= writes a circuit file with either electrical or physical line descriptions for microstrip, stripline, coax or slabline (coupled round rods between flat ground planes).

Table 6-1 *L-C filter structures synthesized by the software program =FILTER=.*

LOWPASS (ALL-POLE)
  Minimum Inductor
  Minimum Capacitor
HIGHPASS (ALL-POLE)
  Minimum Inductor
  Minimum Capacitor
BANDPASS (ALL-POLE)
  Minimum Inductor
  Minimum Capacitor
  Top-C Coupled Parallel Resonator
  Top-L Coupled Parallel Resonator
  Shunt-C Coupled Series Resonator
  Tubular
  Symmetric Transform
  Blinchikoff Flat Delay
BANDSTOP (ALL-POLE)
  Minimum Inductor
LOWPASS (ELLIPTIC)
  Minimum Inductor
  Minimum Capacitor
HIGHPASS (ELLIPTIC)
  Minimum Inductor
  Minimum Capacitor
BANDPASS (ELLIPTIC)
  Conventional Transform
  Zig-Zag Transform
BANDSTOP (ELLIPTIC)
  Minimum Inductor

Table 6-2 *Distributed filter structures synthesized by the software program =M/FILTER=.*

ALL-POLE
  End-Coupled Bandpass
  Edge-Coupled Bandpass (tapped and line-coupled)
  Hairpin Bandpass (tapped and line coupled)
  Combline Bandpass (tapped and line coupled)
  Interdigital Bandpass (tapped and line coupled)
  Stepped-Impedance Bandpass
  Stepped-Impedance Lowpass
  Hybrid Highpass
  Direct-Coupled Bandstop
  Edge-Coupled Bandstop
ELLIPTIC
  Geffe/Admittance Invertor Bandpass
  Direct-Coupled Elliptic Lowpass

As is evident in Figure 6-2c, a simulator program circuit description for a physical implementation consists of the desired

Table 6-3 *Transfer function approximations supported by =FILTER= and =M/FILTER=.*

ALL-POLE APPROXIMATIONS
  Butterworth (singly and doubly terminated)
  Chebyshev (singly and doubly terminated)
  Bessel
  Transitional Gaussian (6 and 12 dB)
  Equiripple Linear Phase (0.05 and 0.5 degree)
  Singly-Equalized
  User g-Values
ELLIPTIC APPROXIMATIONS
  Cauer-Chebyshev
  Elliptic Bessel
  User g-Values

transmission elements plus models for the appropriate discontinuities such as open line ends, via holes to ground, steps in width, bends, tees, and crosses.

6.3   Simulation

Ideally, only synthesis is required. The filter is synthesized and when constructed, the prototype achieves all the desired characteristics. Of course in practice this does not happen. Culprits include mathematical simplifications required for synthesis, component parasitics, distributed element discontinuities, dissipation and radiation losses, dispersion, package modes and measurement error. Prior to the advent of digital computer simulation, the recourse was to build and modify a prototype to achieve the desired filter characteristics. Today, high-speed simulation allows us to perform many of these modifications before construction of the physical prototype.

The responses in Figure 6-1 were computed and displayed using the simulation program =SuperStar= [1] running on an IBM Personal Computer. Ideal L-C models were used for the filter responses displayed in the window on the upper left while microstrip line and discontinuity models were used to compute the responses displayed in the bottom two windows. The responses of the equivalent microstrip form of the filter are given in Figure 6-1 on the lower left. The microstrip responses represent an accurate simulation of an actual filter including the effects of the electrical behavior of the transmission lines, dielectric and conductor loss, dispersion, and discontinuities. Not considered are radiation losses and package modes.

Notice the cutoff frequency is lower than 2250 MHz and the passband return loss and transmission ripple are unequal. The rejection at 4500 MHz is unchanged from the L-C filter case but it must be noted the cutoff frequency is lower.

To what degree do each of the non-ideal characteristics listed above contribute to the discrepancies? Contributions are difficult to isolate using a physical prototype but may be studied in detail using simulation. This greater understanding contributes to a quicker and more accurate correction of the design to achieve the desired objectives.

To obtain a better understanding of the characteristics of this specific filter and to illustrate how computer aided strategies might be applied to any electrical filter, we will use simulation to dissect this example.

6.4 Lumped/Distributed Equivalent Accuracy

First we will assess the accuracy of the equivalence of the lumped and distributed elements used in this filter. On the left in Figure 6-3 the responses of the ideal L-C filter are repeated. On the right are the responses of the stepped-impedance distributed equivalence using ideal electrical transmission line models.

The =SuperStar= circuit file used to generate these responses is listed in Table 6-4. The syntax of this file structure is described in detail in the operation manual and in the Eagleware Technical Overview [2]. The structure is familiar to designers who use simulation programs. The circuit file consists of CIRCUIT, EQUATE, and one or more WINDOW blocks, in this case two. In the top half of the CIRCUIT block the L-C filter is described using capacitor and inductor models. This filter is defined as a two-port labeled IDEAL_LC. In the bottom half of the CIRCUIT block, the stepped impedance lowpass is described using ideal transmission line electrical models and defined as the two-port IDEAL_TL.

The equivalences listed in Figure 3-15 provide an infinite number of distributed equivalences for a given lumped element with line impedance or electrical length being a selectable parameter. In this circuit file the length and impedance of the transmission lines have been defined by equations in the EQUATE block. The first line after the EQUATE block header

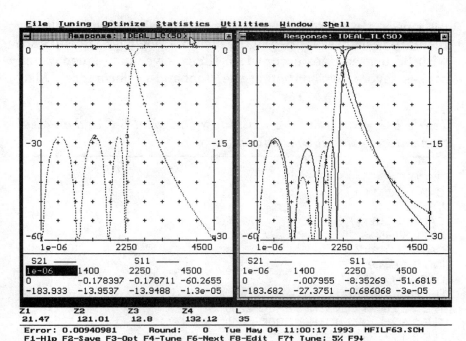

Figure 6-3 *The ideal L-C filter is repeated on the left and an ideal transmission line equivalence with 20 degree lines (solid) and 35 degree lines (dashed) is given on the left.*

defines the initial electrical length as 20 degrees. The "?" preceding the number signifies the electrical length is tunable in =SuperStar=. The next four statements assign values to the line impedances *Z1* through *Z4*. The constants in these expressions are the reactances of the corresponding L-C elements at the cutoff frequency of the lowpass, 2250 MHz. As the electrical length is tuned, the appropriate values of line impedances are computed by these expressions and the resulting values are used in the ideal transmission line filter to compute the responses. The syntax and available functions for the EQUATE block is described in the =SuperStar= Technical Overview [2].

The solid responses on the right in Figure 6-3 are with electrical line lengths of 20 degrees at the cutoff frequency. The dashed responses are with 35 degree lines. The shorter lines clearly result in responses more like the ideal L-C lowpass. Reentrance with 35 degree lines occurs 1.75 times lower in frequency and causes severe degradation of the stopband performance. Also,

Table 6-4 =SuperStar= circuit file for an ideal L-C and an ideal transmission line lowpass filter.

```
CIRCUIT
 CAP 1 0 C=1.89
 IND 1 2 L=4.91
 CAP 2 0 C=3.17
 IND 2 3 L=5.36
 CAP 3 0 C=3.17
 IND 3 4 L=4.91
 CAP 4 0 C=1.89
 DEF2P 1 4 IDEAL_LC
 TLE 1 2 Zo=Z1 LENGTH=L FREQ=2250
 TLE 2 3 Zo=Z2 LENGTH=L FREQ=2250
 TLE 3 4 Zo=Z3 LENGTH=L FREQ=2250
 TLE 4 5 Zo=Z4 LENGTH=L FREQ=2250
 TLE 5 6 Zo=Z3 LENGTH=L FREQ=2250
 TLE 6 7 Zo=Z2 LENGTH=L FREQ=2250
 TLE 7 8 Zo=Z1 LENGTH=L FREQ=2250
 DEF2P 1 8 IDEAL_TL
EQUATE
 L=?20
 Z1=37.43*SIN(L)
 Z2=69.41/SIN(L)
 Z3=22.31*SIN(L)
 Z4=75.78/SIN(L)
WINDOW
 IDEAL_LC(50)
 GPH S21 -60 0
 GPH S11 -30 0
 FREQ
 SWP 0 4500 91
WINDOW
 IDEAL_TL(50)
 GPH S21 -60 0
 GPH S11 -30 0
 FREQ
 SWP 0 4500 91
```

when the lines are 35 degrees long at the cutoff frequency, the tangent function is less linear within the passband, which contributes to passband equiripple degradation. At longer line lengths, a more accurate lumped/distributed equivalent model includes an inductor-capacitor-inductor "tee" network or a capacitor-inductor-capacitor "pi" network to represent a distributed element. We generally avoid these more complex models because the shorter line lengths yield improved filter performance. Also, we are able to correct the errors of the equivalences during computer optimization which is also required to correct other problems.

Why not use very short lines to mitigate these difficulties? The answer is illustrated in Table 6-5 which lists the required line impedances for this stepped-impedance lowpass versus the selected line length. With 15 degree lines, the minimum line impedance required is 6.6 ohms and the maximum line impedance is 232 ohms. This extreme range is difficult if not impossible to realize in microstrip. Line lengths in the vicinity of 25 to 30 degrees are required to minimize realization difficulties in this filter.

6.5   Physical Models

We select 25 degree lines and proceed with the design by consideration of discontinuities, dispersion and losses using physical models in the simulator. The circuit file is listed in Table 6-6. For reference, the ideal L-C filter is defined as the two-port IDEAL_LC. The physical circuit is defined as MSTRIP. The substrate code in the circuit file, SUB, specifies the parameters of the PWB. In this case we are using PTFE board with a dielectric constant of 2.55 and loss tangent of 4×10^{-6}. The metalization is copper with a resistivity relative to copper of 1 and a thickness of 0.71 mils (0.5 oz). The board roughness is 0.01 mils. The remaining description consists of a cascade of microstrip lines (MLI codes) and microstrip step discontinuities (MST codes). Because the filter is symmetrical about the center, half of the the microstrip filter is described and stored in the

Table 6-5 *Required line impedances versus line length to realize the stepped-impedance lowpass filter in Figure 6-2.*

LENGTH (degrees)	Z1	Z2	Z3	Z4 (ohms)
15	9.69	268.2	5.77	292.8
20	12.8	202.9	7.63	221.6
25	15.8	164.2	9.42	179.3
30	18.7	138.8	11.2	151.6
35	21.5	121.0	12.8	132.1
45	26.5	98.2	15.8	107.2
60	32.4	80.1	19.3	87.5

two-port HALF. The entire filter is constructed by cascading HALF with itself with the input and output ports reversed.

The WINDOW section of the circuit file specifies what output parameters to display and the frequency sweep. The optimization block specifies a return loss of 14 dB or better from dc to 2250 MHz (the passband) and at least 31 dB of rejection above 3100 MHz. Optimization is discussed in more detail in a later section.

On the left in Figure 6-4 the responses of the ideal L-C filter are repeated again. On the right, the responses with physical microstrip line and step discontinuity models are displayed. The solid traces on the right are before optimization and the dashed curves after optimization of the physical model. The line widths on the 50 mil dielectric board range from about 18 mils to 800 mils to realize the impedances of the electrical model filter. The line lengths range from 330 to 369 mils for the 25 degree electrical length. The line lengths are unequal because the effective dielectric constant varies slightly with the line width. Because we plan to optimize the line widths to compensate for the discontinuities, we decided to use equal line lengths, 350 mils. The line widths before optimization are indicated in the circuit file and the line widths after optimization are given near the bottom of Figure 6-4.

Table 6-6 *Circuit file with ideal L-C and microstrip implementations of the stepped-impedance lowpass. The microstrip filter is set up for optimization of line widths to recover the response.*

```
CIRCUIT
 CAP 1 0 C=1.89
 IND 1 2 L=4.91
 CAP 2 0 C=3.17
 IND 2 3 L=5.36
 CAP 3 0 C=3.17
 IND 3 4 L=4.91
 CAP 4 0 C=1.89
 DEF2P 1 4 IDEAL_LC
 SUB Er=2.55 Ta=4E-4 Rho=1 Tmet=.71 Rough=.01 Units=.0254
 MLI 1 2 W=W50  L=100          H=50
 MST 2 3 O=SY NARROW=W50 WIDE=W1 H=50
 MLI 3 4 W=W1   L=350          H=50
 MST 5 4 O=SY NARROW=W2  WIDE=W1 H=50
 MLI 5 6 W=W2   L=350          H=50
 MST 6 7 O=SY NARROW=W2  WIDE=W3 H=50
 MLI 7 8 W=W3   L=350          H=50
 MST 9 8 O=SY NARROW=W4  WIDE=W3 H=50
 MLI 9 10 W=W4  L=175          H=50
 DEF2P 1 10 HALF
 HALF 1 2 0
 HALF 3 2 0
 DEF2P 1 3 MSTRIP
EQUATE
 W50=139
 W1=?434
 W2=?23.1
 W3=?797
 W4=?17.8
WINDOW
 IDEAL_LC(50)
 GPH S21 -60 0
 GPH S11 -30 0
 FREQ
 SWP 0 4500 91
WINDOW
 MSTRIP(50)
 GPH S21 -60 0
 GPH S11 -30 0
 FREQ
 SWP 0 4500 91
 OPT
```

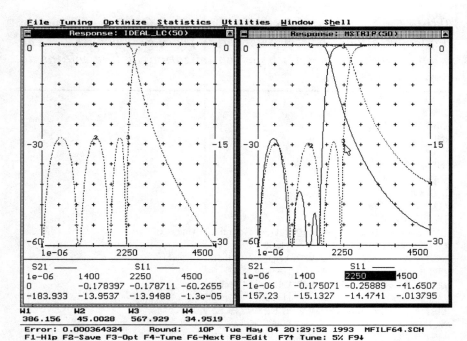

Figure 6-4 *Ideal L-C lowpass on the right and on the right a microstrip lowpass before (solid) and after (dashed) optimization of the line widths.*

Notice that optimization has successfully recovered the passband characteristics but the stopband attenuation of the microstrip filter is degraded due to reentrance. If we were to sweep further into the stopband, we would discover a second passband.

Notice that the line widths after optimization are less extreme than before optimization. This is because the microstrip step discontinuities electrically have a lowpass characteristic. For the stepped-impedance lowpass the steps are self compensating and the difference in the pre- and post-optimized line widths are substantial. It is clear the discontinuities have a profound effect on the responses.

Again repeated for reference, on the left in Figure 6-5 are the responses for the ideal L-C 2.25 GHz lowpass. On the right in Figure 6-5 are the responses of the physical stepped-impedance lowpass, this time with the line lengths optimized to compensate for the discontinuities. The line lengths prior to optimization were all 350 mils and the line lengths after optimization are shown at the bottom of Figure 6-5. With the line widths left at their original extreme values, optimization has compensated for the step discontinuities by shortening the line lengths. The shorter line lengths result in a higher frequency for reentrance which in turn increases the rejection at the upper end of the sweep band. Notice that the rejection in Figure 6-5 at 4500 MHz is approximately 7 dB better than the rejection in Figure 6-4.

On a 33 MHz 80486 Intel CPU based machine the optimization in Figure 6-5 required 11 rounds and 1.8 minutes. In the =M/FILTER= section of this chapter we will discuss discontinuity absorption techniques which substantially reduce optimization requirements, which can save significant time for more complex filter structures.

The dashed traces on the right in Figures 6-4 and 6-5 now represent the responses of the stepped-impedance lowpass considering simulation of the electrical structure, the physical lines including dielectric and conductor losses and dispersion, and analytical models for the step discontinuities. The discrepancies between the simulated and actual responses are generally small provided the range of parameters of the analytical models in the simulator are not exceeded. Numerous examples of simulated and measured responses of various microwave filters are given in the chapters which follow.

6.6 Simulation Technologies

Modern circuit simulators fall into three major categories: linear simulators, SPICE-derived programs, and harmonic balance techniques. Efficient simulators have a few salient features in

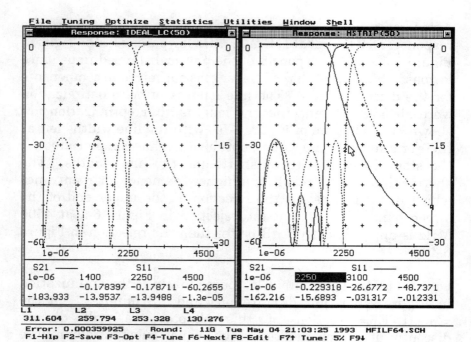

Figure 6-5 *Ideal L-C lowpass on the left and microstrip lowpass on the right before (solid) and after (dashed) optimization of the line lengths.*

common; the circuit descriptions may be stored, recalled and edited, simulation is accurate and the process is general, that is, if the user can describe the network the simulator can compute the response.

Linear simulators, such as =SuperStar=, utilize closed form equations to compute the frequency domain response. SPICE products are almost invariably based on SPICE2 or SPICE3, developed at the University of California, Berkeley, in the early 1970s. SPICE products solve non-linear differential equations for a network using iterative techniques because a closed-form solution is unknown. Both the frequency and time domains are supported.

Harmonic balance provides non-linear analysis, but is restricted to steady state behavior. Each simulator category compromises some desirable attribute. The most demanding design requirements are best satisfied by using more than one simulator class. Some of the advantages and disadvantages of these three simulator classes are listed in Table 6-7.

The disadvantages of SPICE simulation worsen with increasing frequency. Also, time domain data is often of less interest to higher frequency engineers because of the difficulty associated with measuring voltages at high frequencies. Accurate high frequency network analyzers operate in the frequency domain. At lower frequencies, voltage data is less elusive and more insightful. For these reasons, low frequency (below about 10 MHz) engineers typically use SPICE simulators and high frequency engineers use linear simulators. Today, we are seeing more and more cross utilization of these simulator classes and this is healthy. High frequency engineers, when the extra effort of model study and accuracy checking are justified, can benefit from time domain and non-linear simulation. Low frequency engineers can benefit from the real-time tuning and optimization

Table 6-7 *Advantages and disadvantages of three important circuit simulator classes.*

TYPE	ADVANTAGES	DISADVANTAGES
Linear	Highly accurate	Frequency domain only
	Simpler modeling	Linear only
	Very fast & interactive	
	Tuning & optimization	
SPICE	Frequency, time & transient domains	Very slow
	Bias simulation	Elusive models
	Linear & non-linear	Accuracy compromised
		Convergence issues
Harmonic balance	Linear & non-linear	Steady-state only
		Elusive models
		Slow

capabilities of a linear simulator to optimize the frequency domain performance of the circuit.

Harmonic balance simulators grew out of a need to resolve difficulties associated with SPICE simulation; slow execution, lack of convergence and at high frequencies, accuracy and elusive active models. Although helpful, harmonic balance is unfortunately a compromise. Transient and dc simulation are unsupported, and accurate active device modeling is still difficult.

A fourth class of simulation, Volterra-Series, has non-linear capability. It is several times slower than linear nodal simulation, and therefore at least an order of magnitude slower than linear two-port simulation. However, this is fast in relation to other non-linear simulators. Unfortunately, only weakly non-linear circuit simulation is accurate, and the time-domain is unsupported. It is better suited for primarily linear devices, such as class-A amplifiers, than for oscillators and class-C amplifiers.

6.7   Analysis

Analysis is the basic operating mode of simulators. Using analytical models and computational algorithms, the simulator computes and displays the responses of a network described by user entry of a circuit file text list or a schematic. Analysis is a one shot process. Because creating the circuit description requires several minutes, whether program execution occurs in 1 millisecond or 10 seconds is of little consequence.

6.8   Tuning

If execution speed is sufficiently fast, tuning becomes an effective method of modifying and improving the performance of a circuit. What is sufficiently fast? If execution time exceeds a few seconds, activating an editor, modifying a circuit file value and repeating the analysis adds only a modest amount of time.

Tuning degenerates to simple analysis. If the total execution time is a fraction of a second and tuning is accomplished by an equally quick process, simulator tuning becomes as interactive, effective and insightful as tuning a prototype with a fast network analyzer. Total execution times below one second are required to achieve true real-time interactivity. This becomes obvious when tuning is attempted with a network analyzer sweep time exceeding a few seconds.

=SuperStar= utilizes a unique node elimination algorithm, element classes, output parameter classes and model caching to achieve exceptionally fast execution speed. This is further discussed in Sections 6.11 and 6.12.

With =SuperStar=, components marked in the circuit file with a question mark appear on the bottom row of the computer screen, and they are tuned by simply selecting one and tapping on the up or down cursor keys. The tuning step size is controlled by the user. As the component is tuned, the circuit responses are recomputed and displayed in real-time.

6.9   Optimization

When only a few variables in a circuit require adjustment, tuning is an effective tool. As the number of variables increases, visualization of the multi-dimensional variable space is difficult and tuning becomes less effective, so optimization is the preferred tool. A circuit optimization is not "run," but rather "played." Optimization is often a compromise of conflicting requirements with no exact solution. Effective use of optimization consists of an attempt, evaluation of the results, adjustment of the goals and weight, and further optimization.

=SuperStar= includes two distinctly different optimization algorithms; a gradient search and a pattern search with adaptive and independent step size for each variable. While these routines are proprietary, they are described here in sufficient detail for effective use of the optimizer. Also, it cannot be overemphasized that a major factor contributing to the effectiveness optimization is execution speed.

Gradient optimization is effective in the early phase of an optimization effort. It is reasonably tolerant of poor initial component values and a large number of components. It often makes significant progress after only the number of rounds equal to the number of variables being optimized. However, gradient search algorithm progress tends to halt before achieving optimum final values.

The search is effective in the final phases of an effort [3]. The current =SuperStar= routines improve the published routines because 1) adaptive and independent variable step size was introduced and 2) fewer evaluations of the circuit are required for a given number of steps of the variables. The pattern search algorithm is very resistant to "hang."

=SuperStar= contains an automatic mode which initially invokes the gradient optimizer. When progress halts, as evidenced by a suspension in the decline of the error from target (objective function), the pattern search algorithm is invoked. A fixed number of pattern searches is applied and then the gradient optimizer is again invoked.

Each set of component values results in an error from the desired response. The error per frequency is given by

$$Err = \left(W_{mn}(c_{mn} - C_{mn})^p + U_{mn}(p_{mn} - P_{mn})^p + WNF(nf - NF)^p \right.$$
$$\left. + WDL(dly - DLY)^p + WKS(kst - KST)^p \right)^{0.5} \tag{1}$$

where

c_{mn}, C_{mn}	=target and actual linear S-parameters
W_{mn}	= linear S-parameter magnitude weight factor
p_{mn}, P_{mn}	= target and actual S-parameter phase
U_{mn}	= S-parameter phase weight factor

nf, NF	= target and actual linear noise figure
WNF	= noise figure weight factor
dly, DLY	= target and actual transmission group delay
WDL	= group delay weight factor
kst, KST	= target and actual "K" stability factor
WKS	= stability factor weight factor

The exponent "p" is always even, therefore the magnitude of each error contribution is positive. When the user selects pattern search, $p=2$, which results in a root-mean-squared error minimization. When the user selects gradient or automatic optimization, p equals 6 which results in a Chebyshev error minimization.

Each line in an OPT block in the circuit file adds to the error value as determined by the above equation. The total error value is the sum of the errors per frequency divided by the number of frequencies, these added for each OPT block line. If a parameter has not been specified in a line of the OPT block, its weight is zero. A specified parameter has a default weight of 1 unless modified by the weight option. The weight for all parameters is zero for frequencies outside the frequency range of a given OPT line.

The optimization routine attempts to reduce the total error value by adjusting the values of all components in the circuit file marked with a "?." The error and number of rounds are displayed during the optimization process. A "P" suffix on the number of rounds signifies the pattern search is currently active, while a "G" indicates gradient optimization is active. Each round evaluates all marked components.

If the user selects gradient optimization or the automatic mode from the menu, optimization begins immediately. Optimization continues until interrupted by pressing Esc or until the error reaches zero. Optimization may be interrupted and restarted at will. Manual tuning may be applied during the interruption. If the user selects pattern search optimization, the variable step size is prompted and optimization begins. For broadband circuits, a moderate step size such as 5% is reasonable. For narrow-band circuits a smaller initial step size is recommended. Because the step size is adjusted dynamically during optimization, the initial step size is not critical.

After optimization has run a while, variable step sizes normally decrease. If optimization is interrupted to manually adjust variables, it is good practice to specify a smaller step size when restarting optimization.

If too large an initial step size is chosen, the early rounds of optimization do not modify the circuit values; they are used to reduce the variable step sizes. On occasion, the error value may actually increase. This attribute of =SuperStar= optimization allows it to "wander" in search of a better ultimate solution. If this happens at the beginning of a run, it may be indicative of too large an initial step size.

A line in the OPT block might be

10 70 S21>−1

Each line in the OPT block begins with two numbers which indicate the frequency range that applies to the conditions of that line. In this example, the forward gain, S_{21}, is to be greater than −1 dB over the frequency range of 10 to 70 MHz. This OPT line is very simple. Each line may specify one or more of the following parameters

S11, S21, S12, S22   Decibel S-parameter magnitudes
P11, P21, P12, P22   S-parameter phase in degrees
DLY   Group delay in nS
KST   Stability factor "K"
NFD   Decibel noise figure

The maximum number of lines in an OPT block is limited only by available memory. The allowed operators are =,>,< and %.

The = operator attempts to optimize to the specified value. < or > attempt to optimize a parameter to be less than or greater than the specified value. The % operator attempts to flatten the specified parameter without regard for a specific value.

This three line optimization block for a bandpass filter

```
10 40 S21<-40
55 85 S21>-1 DLY%
100 130 S21<-40
```

attempts to achieve at least 40 dB of rejection in lower and upper stopbands and less than 1 dB insertion loss with flat delay in the passband.

This two line optimization block for an amplifier

```
2000 4000 S21>11.5 S11<-10 S22<-10
2000 4000 S21<12.5
```

attempts to achieve better than 10 dB of return loss in an amplifier with 11.5 to 12.5 dB of gain. A similar optimization block would be

```
2000 4000 S21=12 S11<-10 S22<-10
```

except the two previous lines specify that any gain between 11.5 and 12.5 dB is acceptable. Optimization could concentrate more on achieving the return loss. Additional examples of optimization are given in Figures 6-4 and 6-5, and throughout the remaining chapters.

6.10   Statistical Analysis

Production oriented design considers the effects of component tolerances on circuit performance to gain confidence that the yield will fall within acceptable limits.

One method of gaining confidence is to consider worse case scenarios. The circuit response is computed with each component stepped up or down in value by the appropriate tolerances. The response is observed while all variables are stepped in the direction resulting in the worst possible outcome for the parameter being considered. This process is fast and insightful with a real-time simulator such as =SuperStar=. However, the outcome is generally pessimistic. Redesign to insure worse case scenarios meet specifications often results in greater cost than rejecting or repairing a few units which fail test.

Monte Carlo analysis evaluates circuit behavior for a sample run size with a random distribution of component values within specified limits. It is a statistical process. It does not predict with certainty what will happen, but it identifies likely production performance ranges.

Consider the filter responses in Figure 6-6. On the left are the transmission and input return loss responses plotted on a rectangular grid and on the right is the input return loss plotted on a Smith chart for a 25 sample run of the microstrip lowpass with a tolerance on the dielectric constant of ±0.05, line widths of ±1 mils and substrate (board) thickness of ±0.5 mils. The circuit file is given in Table 6-8. Question marks preceding component values in the file indicate those values are included in the Monte Carlo analysis. Notice a variable, W_{etch}, is introduced with a nominal value of 1 mil. This is done because a tolerance in line widths is likely to apply to all lines because of over or under etching. W_{etch} is given a tolerance of ±100% representing a line width tolerance of ±1 mil.

The responses are painted for a distribution of component values around the nominal values. The process continues until the specified number of runs (sample size) is achieved. In this case, 23 of the 25 samples (92%) satisfy the target specifications given in the YIELD section of the WINDOW block. At the end of the run, the markers read response values for the nominal

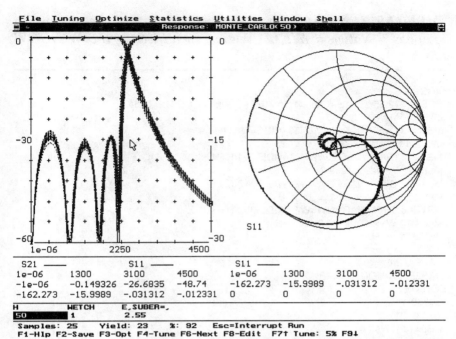

Figure 6-6 *Monte Carlo response paints for the microstrip lowpass with tolerance on the board thickness, dielectric constant and transmission line widths.*

component values.

In this example, the effects of tolerances on the selected parameters are reasonably minimal and a yield specification relatively close to the nominal responses still results in a yield of 92%. We will discover in later examples that such successful outcomes are not universal. Monte Carlo, like tuning, can be insightful and provides an understanding of circuit behavior and problem areas.

As we have seen, the circuit file for a Monte Carlo run is identical to a circuit file for analysis, tuning or optimization, except for a YIELD section which specifies response limits for a successful unit. The YIELD section is identical to an OPT section except the "=" and "%" operators are rarely used because

Table 6-8 *Circuit file for a Monte Carlo run of the microstrip lowpass. Notice a YIELD section is added to the WINDOW block.*

```
CIRCUIT
 SUB Er=?2.55 Ta=4E-4 Rho=1 Tmet=.71 Rough=.01 Units=.0254
 MLI 1 2 W=W50  L=100           H=H
 MST 2 3 O=SY NARROW=W50 WIDE=W1 H=H
 MLI 3 4 W=W1   L=311.6         H=H
 MST 5 4 O=SY  NARROW=W2 WIDE=W1 H=H
 MLI 5 6 W=W2   L=259.8         H=H
 MST 6 7 O=SY NARROW=W2  WIDE=W3 H=H
 MLI 7 8 W=W3   L=253.3         H=H
 MST 9 8 O=SY NARROW=W4  WIDE=W3 H=H
 MLI 9 10 W=W4  L=130.3         H=H
 DEF2P 1 10 HALF
 HALF 1 2 0
 HALF 3 2 0
 DEF2P 1 3 MONTE_CARLO
EQUATE
 H=?50
 Wetch=?1
 W50=138+Wetch
 W1=433+Wetch
 W2=22.1+Wetch
 W3=796+Wetch
 W4=16.8+Wetch
WINDOW
 MONTE_CARLO(50)
 GPH S21 -60 0
 GPH S11 -30 0
 SMH S11
 FREQ
 SWP 0 4500 91
 YIELD
 0 2200    S11<-13.5
 4000 4500 S21<-42
```

the yield would be zero. The ">" and "<" operators are used to specify the range of output parameters which constitute a successful unit. The OPT section is used to find component values which result in the desired nominal responses. The YIELD section is used to set acceptable limits for definition of what is a successful unit during Monte Carlo analysis. If the

YIELD section is absent, Monte Carlo will report a yield of 100%. The OPT section is not used by Monte Carlo.

The sample size, random number seed and component distributions are changed in the Setup Monte Carlo section of the Statistics menu. The random numbers used for component distribution are derived from the specified seed. An integer seed between −32768 and +32767 is specified. Runs with the same seed, circuit file and sample size are identical. This provides the user with both the ability to repeat a specific run or to create 65,536 different runs of a specified sample size.

In the previous Monte Carlo run we accepted the default seed, 0, and sample size, 25. A uniform distribution for all three parameters was selected. A uniform distribution signifies that all parameter values over the specified range are equally probable. The percentage above and below nominal may be independently specified. For example, −80% to +20% (common with high dielectric-constant bypass capacitors) is an acceptable specification.

A number of Set Variable Stats boxes appear in sequence, once for each component value marked with a "?." At the top of each Set Variable Stats box is the component identifier and the nominal component value. Next are input fields for the component %up and %down range (or %one sigma). Because the substrate dielectric constant in this example was 2.55 ±0.05, a percentage range of ±1.96 was entered. The uniform distribution radio button was selected.

An alternative to uniform distribution is a normal distribution. The normal distribution results when a large number of independent events produces additive effects. The distribution curve is bell shaped around the nominal value. The sum of several tossed dice follows a normal distribution for repeated tries. A continuous normal distribution is approximated in =SuperStar= as the sum of ten independent events, each with 65,536 equally probable outcomes.

The mean value is the nominal value specified in the circuit file as a constant preceded by a "?." The user specifies the one sigma deviation. Approximately 68.3% of component values fall within the one sigma limit. Approximately 99.7% of component values fall within three sigma limits. A significant number of values exceed one sigma deviation. Components outside three sigma limits are rare.

Activating Monte Carlo automatically creates a file with the circuit filename and the extension ".MC" if such a file does not already exist. This ".MC" file specifies a ±5% uniform distribution for all variables, a seed of 0, and a sample size of 25. User inputs in Monte Carlo set-up boxes overwrite the default specifications. This user-created ".MC" file remains in effect for that circuit file unless the ".MC" file is erased.

As the number of components affecting the response increases, it becomes unlikely that a run will exist where all values fall at extreme values in the direction causing the worst response. However, when only a few components affect the response, Monte Carlo is more likely to produce a near worse case run. Also, if one component exhibits the greatest sensitivity, a near worse case run is more probable.

When a number of response specifications exist, more than one set of worse case component values may exist. The set of component directions are typically different for different response specifications. Although Monte Carlo may not find worst case responses, the user is relieved of the tedium of manually tuning several sets of worse case component values.

The error value reported during optimization, tuning and Monte Carlo are all computed by the same algorithm, and therefore relate directly to each other.

Selecting Sensitivity from the Statistics menu displays sensitivity plots in sequence for each component value marked with a "?" in the circuit file. Display pauses for viewing for each variable until the user strikes the <Enter> key. In each sensitivity plot, the responses are displayed with components at the nominal and specified deviation up and down values. The deviation values are the limits for uniform distribution, and the one-sigma values for normal distribution.

Sensitivity analysis is useful for characterizing and identifying individual relationships between components and the circuit responses. It is yet another tool to assist the designer with understanding circuit behavior and managing production yield.

Selecting Generate Report in the General Setup box of the Setup Monte Carlo item in the Statistics menu will write an ASCII file to disk which contains information on each sample in Monte Carlo run.

Monte Carlo analysis is statistical. Results will vary from run to run. For example, if the above run is repeated with a seed other than zero, the responses and yield may be different.

6.11 Node Elimination Algorithm

Node elimination is a network reduction process developed for the =SuperStar= simulator. Node elimination is a novel approach based loosely on both the older chain matrix approach and modern graph theory. Node elimination offers several improvements over existing chain matrix techniques.

Easy entry of circuit parameters and topology
Unrestricted circuit topologies
When combined with specialized element classes, even faster execution that conventional chain matrix algorithms.

Sparse matrix systems are touted as computationally efficient. While these systems are more efficient than standard matrix inversion techniques, they are not as efficient as node elimination. Table 6-9 shows a comparison of the number of calculations required for analyses of the simple network in

Table 6-9 *Number of floating-point operations required for analysis of the circuit in Figure 6-7.*

REQUIRED PROCESS	OPERATIONS	
	PER ELEMENT	TOTAL
NODE ELIMINATION		
Calculate admittance/impedance (finite Q)	2	6
Add admittances		2
Convert admittance to impedance and add		8
Total number of operations per frequency		16
SPARSE MATRIX		
Zero out needed sparse matrix cells (3x3 case)		7
Calculate admittances (finite Q0		6
Add Y-matrices to sparse matrix	8	24
Invert and solve sparse matrix (variances)		220
Total number of operations per frequency		251

Figure 6-7 using =SuperStar= node elimination and sparse matrix techniques available in other circuit simulators.

Node elimination achieves this efficiency by combining elements or groups of elements much like chain matrix techniques, thus removing (or eliminating) nodes from the circuit. For example, when analyzing the structure in Figure 6-7, the impedances or admittances for each element are calculated. Next, the admittances for the parallel inductor and capacitor are added together. Finally, the impedance of the parallel combination is

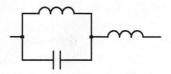

Figure 6-7 *Simple network used to illustrate the effectiveness of the node elimination algorithm.*

added to the impedance of the series inductor. In contrast, sparse and standard matrix techniques

> Create and zero-out a 3x3 standard matrix or sparse matrix list
> Calculate admittances for each element
> Create a 2x2 Y-matrix for each element
> Add the 2x2 Y-matrices into matrix cells (might be combined with the previous step in better simulators)
> Solve the matrix via inversion.

Although simple, this network demonstrates the power of the node elimination technique. For larger circuits, the approximate number of calculations per frequency is given in Table 6-10. Notice that the number of calculations typically required with sparse matrix increases with the square of the part count. This is much better than standard matrix techniques where the number of calculations is proportional to the cube of the part count, but is much poorer than node elimination where the number of calculations increases linearly with the part count. For larger circuits, node elimination efficiency increases relative to any matrix technique.

These figures do not include overhead such as transcendental functions for transmission line calculations and conversion of final results to S-parameters for display since all algorithms would have about the same amount of overhead for these calculations. They also do not include screen and mouse overhead for user interface. High resolution mode video display time becomes a significant fraction of the total time when fast network reduction algorithms are used. For this reason, video accelerator cards are recommended when using =SuperStar= .

6.12 Element & Output Classes

The node elimination routines use six different element classes to drastically reduce the number of calculations required to connect any two elements. For example, if a series impedance is cascaded with an $ABCD$ matrix, the resulting $ABCD$ network

Table 6-10 *Approximate number of floating-point operations for a general network with n total elements, n_p of which are passive elements and n_a are active two-ports.*

REQUIRED PROCESS	FLOPS
NODE ELIMINATION	
Calculate admittance & impedance	$2n_p$
Connect elements	
passive elements maximum	$16(n_p-1)$
passive elements minimum	$2(n_p-1)$
passive elements typical	$10(n_p-1)$
active elements maximum	$110n_a$
active elements minimum	$2n_a$
active elements typical	$\underline{56n_a}$
Total for node elimination, typical	$12n_p+56n_a-10$
SPARSE MATRIX	
Zero out matrix	1*cells
Calculate admittances	$2n_p$
Add Y-matrices to cells	
passive	$8n_p$
active	$18n_a$
Invert and solve sparse matrix	$\underline{8*nodes*cells+52(nodes-2)}$
Total for sparse matrix, typical	$10n_p+18n_a+52*nodes-104$
	$+(8*nodes+1)*cells$
Simplified formula for large circuit	$10n^2+10n_p+18n_a+28.5n-104$

can be determined using only 16 calculations. If the series impedance had been stored as a full *ABCD* matrix, 56 calculations would have been required for the cascade. =SuperStar= element classes include: series impedance, shunt admittance, two types of L-sections, passive *ABCD* matrix, and active *ABCD* matrix. Specialized classes are included for coupled transmission lines, mutual inductors, op-amps, and tee and cross junction discontinuities.

Output classes are established independently for each output window. For example, if noise figure is not requested, it is not calculated. A more subtle refinement is that if the phase of an output parameter is not required, then it is not calculated. When only the magnitude of S_{21} is requested, one arctangent calculation per frequency is saved. Arctangents are CPU intensive functions, requiring the equivalence of 30-50 standard floating point calculations on a numeric processor.

The element and output classes generated by =SuperStar= are invisible to the user. Nonetheless, the end result is the same: =SuperStar= execution speed is drastically increased.

6.13 Detailed CAE Example

Next the design of a microstrip elliptic lowpass filter is explored from start to finish using the =M/FILTER= synthesis program and the =SuperStar= simulation program. We will begin with a filter specification and end with artwork and measured data. This is an introduction to the process used for other filter types in later chapters.

The required lowpass filter has a maximum 0.0432 dB passband ripple through 3300 MHz and must have at least 45 dB of rejection by 3700 MHz. It is constructed on 25 mil thick PTFE board with a dielectric constant of 2.55 and 1 ounce copper. The loss tangent of the board is 0.0004 and the surface roughness is 0.06 mils.

The N-Help section of =M/FILTER= in the Utilities menu is used to determine the filter order. To provide a safety margin, the required rejection is assumed to be 50 dB. The N-Help routine indicates a Chebyshev all-pole transfer approximation requires a filter order of almost 18 while a Cauer-Chebyshev elliptic approximation only requires just over eight. The N-Help screen is given in Figure 6-8. We will design the filter with 8th order for an additional safety margin.

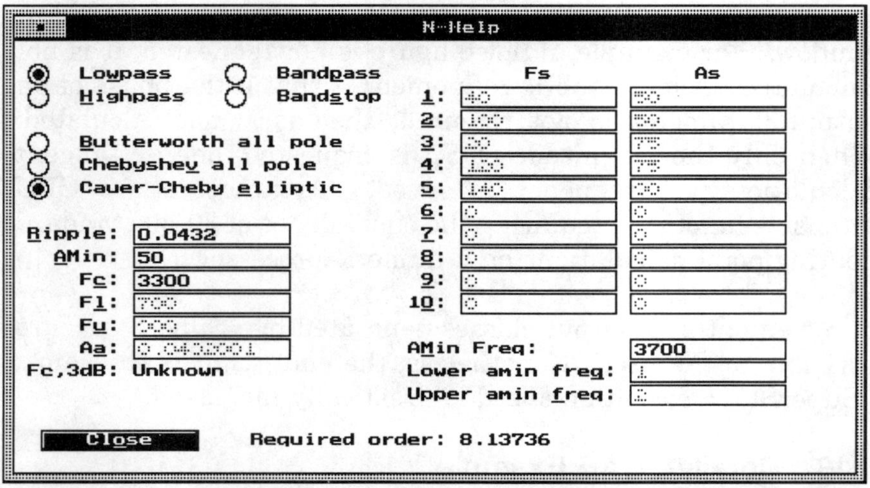

Figure 6-8 N-Help screen from =M/FILTER= which assists with the determination of the required filter order.

The New Start button on the upper left of the =M/FILTER= screen is then selected and the Filter Topology selection window shown in Figure 6-9 appears. The listed filter topologies are structures synthesized by =M/FILTER=. More detailed

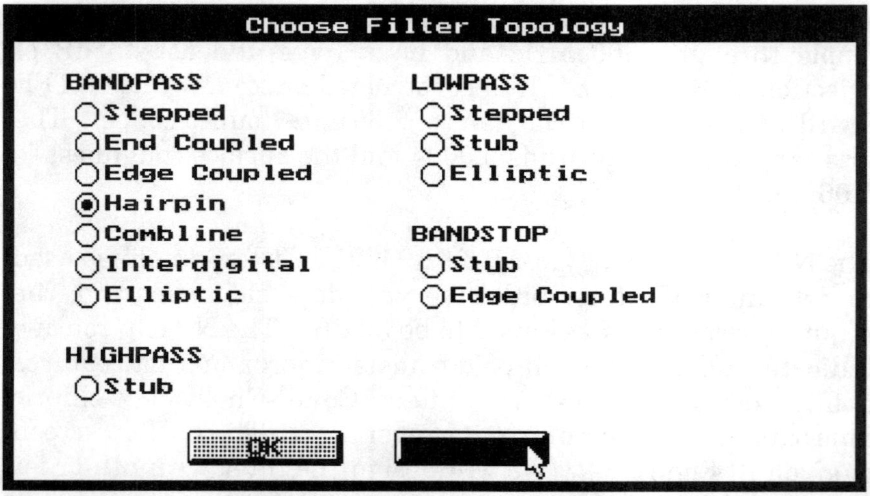

Figure 6-9 =M/FILTER= program Filter Type selection window.

descriptions of these filter topologies, practical issues associated with their use, =SuperStar= simulations of these filters and measured data are given in later chapters.

The elliptic lowpass topology is selected and =M/FILTER= then displays the Filter Shape selection window. If an all-pole topology is selected then all-pole transfer approximation shapes are listed. In this case, an elliptic topology was selected and alternative elliptic transfer shapes are listed. We select the Cauer-Chebyshev with equal input and output termination resistance. Next the Filter Process selection window is displayed and microstrip is selected.

The main =M/FILTER= screen then displays. This screen, after the desired parameters for this filter have been entered is shown in Figure 6-10. On the left in Figure 6-10 is a diagram of the procedures managed by the =M/FILTER= program. =M/FILTER= contains synthesis algorithms and it also manages the entire Eagleware software environment, including simulation, involved in the design of distributed filters.

The procedure diagram indicates the procedure status and serves as selection button to change the procedure. New Start prepares =M/FILTER= for a new design and invokes in sequence the Filter Topology selection window, the Filter Shape window and the manufacturing process window.

After selections in these windows are completed, the Interactive status button and the Calc button are highlighted. The highlighted Interactive button signifies that the user may enter new values in the filter parameter input cells or may select other procedure buttons on the diagram. The filter parameter entry sections at the bottom of the screen initially contain the last user entries. Filter descriptions may be stored and recalled in *.MF$ files accessed from the File menu of =M/FILTER=. When the Calc button is highlighted it signifies that a filter parameter has been changed but the layout has not yet been recalculated and updated on the display. Selecting the Calc

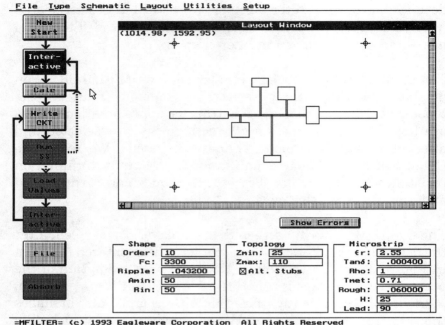

Figure 6-10 =M/FILTER= program main screen. The user enters filter parameters and the program computes and displays the layout of the synthesized filter.

button recalculates and redisplays the layout and extinguishes the Calc button highlight.

For this example, we have entered an order of 10, a lowpass cutoff frequency of 3300 MHz, a passband ripple of 0.0432 dB, minimum attenuation in the stopband of 50 dB and 50 ohm termination resistances. The minimum line impedance we wish to use is 25 ohms and the maximum line impedance is 110 ohms. The parameters of the substrate material are entered on the lower right. The units of metalization thickness, surface roughness, dielectric thickness (H) and the length of leads on the ends are mils, millimeters or custom as setup in the Units section of the Setup menu. In this case the units are mils, thousandths of an inch.

The elliptic lowpass topology is synthesized by first computing the prototype g-values of a Cauer-Chebyshev transfer approximation with the specified A_{min} and passband ripple. The g-values are then scaled to the desired termination resistance and cutoff frequency. Then L-C/transmission line equivalences are used to find the required line lengths given the selected values of Z_{min} and Z_{max}. The series narrow lines through the axis of the filter are derived from the series inductors in the prototype and the stepped-impedance stubs are derived from the series L-C branches to ground in the prototype. The low-impedance line at the output of the filter is derived from the single shunt-capacitor present in even order Cauer-Chebyshev prototypes.

The cross hairs displayed in the layout window mark the edges of the PWB and serve as a via hole reference if required. The position of the cross hairs and other layout details are setup in the Cross Hair Setup section of the Setup menu as shown in Figure 6-11. Check boxes provide for selection of those cross hairs which are to appear on the layout. The cross size specifies the physical size of each cross hair. The top margin is the physical distance from the top cross hairs to the upper-most extent of the filter. The left margin is the physical distance from the left cross hairs to the left-most extent of the filter including the leader. A negative margin causes the leader to extend beyond the cross hairs so a shear at the cross hairs insures the leader extends to the board edge. The width and height specify the horizontal and vertical cross hair spacings. When checked, the Auto Size box causes the cross hair width and height to automatically change to match the filter using the specified margins on all sides of the filter. Layout rotation provides for angular rotation of the layout. This is convenient for slight rotation of edge-coupled filters so the input and output lay on a horizontal axis.

The displayed numeric coordinates (horizontal, vertical) displayed on the upper left in Figure 6-10 are initialized by placing the mouse cursor at the desired reference point and

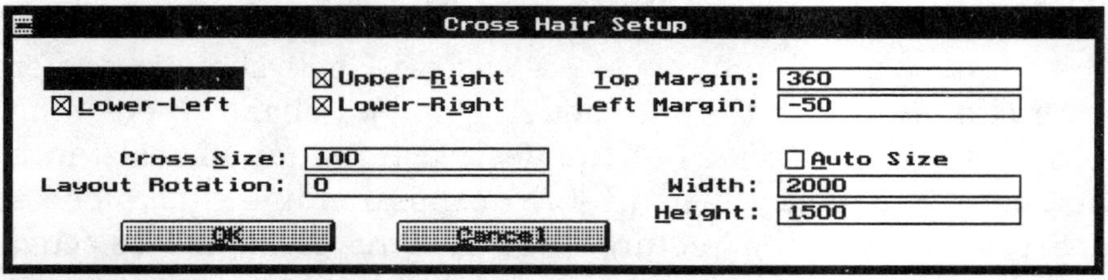

Figure 6-11 *Cross Hair Setup window in =M/FILTER= used to enter layout related parameters.*

clicking the left mouse button. The coordinates with respect to the selected reference point are then displayed. Holding down the Ctrl key and tapping the arrow keys moves around in the layout window. Holding down PgUp key and PgDn key zooms the view in and out for close inspection of layout dimensions using the coordinate display.

6.14 The Next Step: Simulation

The next procedure is simulation of the proposed filter. Writing a circuit file from =M/FILTER= is invoked by selecting the Write CKT procedure button. Two types of =SuperStar= simulator circuit files are written; a generic description using electrical transmission line models and a physical description using models for the selected manufacturing process, such as stripline, microstrip, slabline or coaxial. Either of these two types may be written in a text net-list format (*.CKT) or a schematic format (*.SCH) for the Eagleware =SCHEMAX= program. The electrical model descriptions execute much faster and do not include discontinuities. Although an electrical model simulation is typically not required, for very complex filter structures this preliminary step can save significant optimization time because they execute faster than physical model simulations. The initial optimization is performed with electrical models and a final optimization is performed with slower physical models.

First we select Write Electrical CKT File. The circuit file written by =M/FILTER= is given in Table 6-11. The parameters of the electrical transmission line models, TLE, used in this circuit file are the impedance, Z, the electrical length, L, and the reference frequency, F, for the electrical length. ZI and LI are the input lead line impedance and length in degrees and ZOUT and LOUT are the for the output lead line. Zhi and Zlo are the maximum and minimum line impedances. The values are specified with variables defined in the EQUATE block.

The transmission and return loss responses as computed and displayed by =SuperStar= are given in Figure 6-12. The solid traces are for the filter as synthesized by =M/FILTER=. The

Table 6-11 *=SuperStar= circuit file with electrical transmission line models for the elliptic lowpass written by =M/FILTER=.*

```
'   FILE: MFILF67.CKT              L1=?13.20206
'   TYPE: Elliptic -- Lowpass      L2=?5.723204
'   Fc: 3300 MHz                   L3=?26.54228
' PROCESS: Electrical Description  L4=?28.36878
                                   L5=?32.71231
CIRCUIT                            L6=?13.65291
TLE 1 2 Z=ZI L=LI F=Fc            L7=?21.55297
TLE 2 3 Z=Zhi L=L1 F=Fc           L8=?46.68899
TLE 3 93 Z=Zhi L=L2 F=Fc          L9=?8.651089
TLE 93 94 Z=Zlo L=L3 F=Fc         L10=?24.02533
TLE 3 4 Z=Zhi L=L4 F=Fc           L11=?16.244
TLE 4 101 Z=Zhi L=L5 F=Fc         L12=?21.2398
TLE 101 102 Z=Zlo L=L6 F=Fc       L13=?24.87482
TLE 4 5 Z=Zhi L=L7 F=Fc           L14=?21.26554
TLE 5 109 Z=Zhi L=L8 F=Fc         ZOUT=50
TLE 109 110 Z=Zlo L=L9 F=Fc       LOUT=90
TLE 5 6 Z=Zhi L=L10 F=Fc          WINDOW
TLE 6 117 Z=Zhi L=L11 F=Fc        FILTER(50,50)
TLE 117 118 Z=Zlo L=L12 F=Fc      GPH S21 -5 5
TLE 6 7 Z=Zhi L=L13 F=Fc          GPH S21 -100 0
TLE 7 8 Z=Zlo L=L14 F=Fc          GPH S11 -30 0
TLE 8 9 Z=ZOUT L=LOUT F=Fc        FREQ
DEF2P 1 9 FILTER                  SWP 0 6600 67
EQUATE                            OPT
Fc=3300                           0 3300 S11<-100
ZI=50                             3800 6600 S21<-50 W21=1000
LI=90
Zhi=110
Zlo=25
```

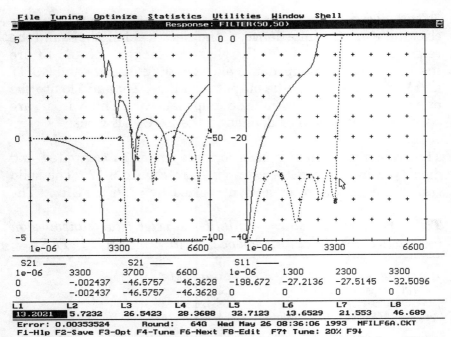

Figure 6-12 *Responses of an electrical model of the elliptic lowpass filter before (solid) and after optimization to recover the responses (dashed).*

cutoff frequency is too low and the passband is far from equal ripple as is evidenced by the return loss. Also the stopband response is not equiripple because the transmission zeros are too low in frequency. The discrepancies are due to failure of the L-C/distributed equivalences. If more extreme high and low line impedances had been used the discrepancies would have been reduced but physical realization would be more difficult.

The dashed traces are after optimization of the line lengths to attempt to recover the response. The first step was to manually shorten all line lengths by steeping them down 20% using the tune mode in =SuperStar=. Next, optimization was invoked using the OPT section in the WINDOW block as written by =M/FILTER=. After a few rounds it was discovered that the stopband attenuation was poor, so a weight factor of W21=1000 was added to the OPT section and optimization was restarted. The result is the dashed traces in Figure 6-12. The electrical line lengths in Table 6-11 are the values after optimization.

Next we investigate the effects of a physical implementation of this filter including microstrip lines, steps and tee discontinuities. Given in Figure 6-13 are the responses for this filter computed from a physical description created by modifying the electrical file. Notice the corner rolloff associated with loses in the microstrip. Also notice the corner frequency is ten percent low and the return loss response is modified, probably because of the step discontinuities. The =SuperStar= circuit file used to compute these responses is given in Table 6-12.

At this point we could optimize the physical file line lengths to attempt to compensate for the effects of the discontinuities.

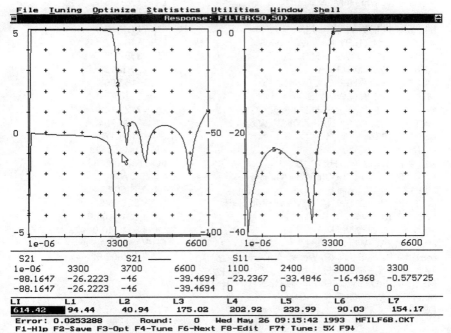

Figure 6-13 *Responses of a microstrip implementation of the elliptic lowpass including microstrip line, step and tee models.*

Instead, we will use available algorithms in =M/FILTER= which automatically absorb the step and tee discontinuities by modifying the length of adjacent microstrip lines. We do this by

Table 6-12 =SuperStar= elliptic lowpass circuit file with a physical description created by converting the electrical model circuit file.

```
CIRCUIT                                         EQUATE
SUB ER=2.55 TAND=.0004 RHO=1 TM=.71             H=25
& ROUGH=.06 UNITS=.0254                         VIAR=20
MLI 1 2 W=WI H=H L=LI                            VIAT=0.71
MST 2 3 O=SY NAR=WI W=Whi H=H                    WI=69.0148
MLI 3 4 W=Whi H=H L=L1                           LI=?610.062
MTE 4 5 100 WT=Whi WS=Whi H=H                    Whi=14.5375
MLI 100 101 W=Whi H=H L=L2                       Wlo=179.206
MST 101 102 O=SY NAR=Whi W=Wlo H=H              L1=?89.0708
MLI 102 103 W=Wlo H=H L=L3                       L2=?43.1034
MEN 103 0 W=Wlo H=H                              L3=?156.106
MLI 5 6 W=Whi H=H L=L4                           L4=?195.026
MTE 6 7 110 WT=Whi WS=Whi H=H                    L5=?236.159
MLI 110 111 W=Whi H=H L=L5                       L6=?71.1138
MST 111 112 O=SY NAR=Whi W=Wlo H=H              L7=?146.272
MLI 112 113 W=Wlo H=H L=L6                       L8=?336.135
MEN 113 0 W=Wlo H=H                              L9=?38.1318
MLI 7 8 W=Whi H=H L=L7                           L10=?163.957
MTE 8 9 120 WT=Whi WS=Whi H=H                    L11=?118.36
MLI 120 121 W=Whi H=H L=L8                       L12=?121.142
MST 121 122 O=SY NAR=Whi W=Wlo H=H              L13=?164.734
MLI 122 123 W=Wlo H=H L=L9                       L14=?135.172
MEN 123 0 W=Wlo H=H                              WOUT=69.0148
MLI 9 10 W=Whi H=H L=L10                         LOUT=?602.863
MTE 10 11 130 WT=Whi WS=Whi H=H                  WINDOW
MLI 130 131 W=Whi H=H L=L11                      FILTER(50,50)
MST 131 132 O=SY NAR=Whi W=Wlo H=H              GPH S21 -5 5
MLI 132 133 W=Wlo H=H L=L12                      GPH S21 -100 0
MEN 133 0 W=Wlo H=H                              GPH S11 -40 0
MLI 11 12 W=Whi H=H L=L13                        FREQ
MST 12 13 O=SY NAR=Whi W=Wlo H=H                SWP 0 6600 67
MLI 13 14 W=Wlo H=H L=L14
MST 14 15 O=SY NAR=Wlo W=WOUT H=H
MLI 15 16 W=WOUT H=H L=LOUT
DEF2P 1 16 FILTER
```

returning to =M/FILTER= from the Shell menu of =SuperStar=. =M/FILTER= automatically maintains a record of the procedure status for each filter design launched from =M/FILTER=. When we return to =M/FILTER= the message displayed in Figure 6-14 appears. =M/FILTER= provides the option of loading the values which have been tuned or optimized in =SuperStar= back into the =M/FILTER= program. This is important for two reasons; =M/FILTER= algorithms use the optimized electrical line lengths and automatically create a physical file including discontinuity absorption and values which have been reloaded are available for the layout plots provided by =M/FILTER=.

Therefore, the Load Values procedure button is selected. The Absorb button momentarily highlights indicating that discontinuities are being absorbed in the adjacent line lengths.

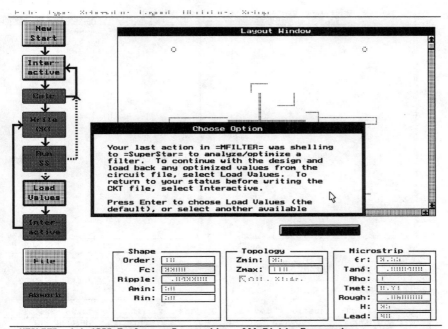

=MFILTER= (c) 1993 Eagleware Corporation All Rights Reserved

Figure 6-14 *Notification message which appears in =M/FILTER= after returning from the first simulation/optimization session in =SuperStar=.*

At this point =M/FILTER= is in a limited interactive mode and access to certain functions is restricted because the original synthesis has been updated by user tuning or optimization. The limited interactive mode allows access to certain functions such as cross hair setup, plotting and file maintenance.

For the next step, selecting Write Physical CKT File causes =M/FILTER= to write a physical model =SuperStar= circuit file using the optimized electrical line lengths and the absorbed discontinuities. The resulting responses are given in Figure 6-15. Notice that even without optimization, the physical model discontinuity absorption by =M/FILTER= results in a microstrip filter with responses almost identical to the corrected electrical model lowpass in Figure 6-12. A final physical model =SuperStar= circuit file is given in Table 6-13.

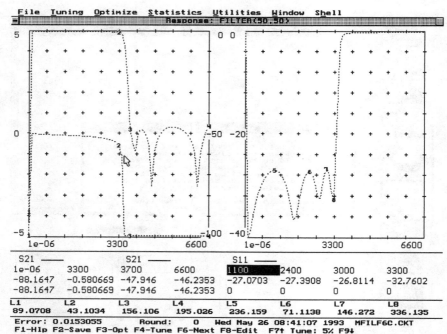

Figure 6-15 *Responses of the final microstrip elliptic lowpass created using =M/FILTER= synthesis and =SuperStar= simulation.*

Now we generate artwork for this filter by selecting Plot Layout from the Layout menu of =M/FILTER=. The =M/FILTER= Layout Print/Plot Setup window shown in Figure 6-16 appears. There are three layers (colors) in the plot; metalization, via holes and cross hairs. Each layer may be on or off for the plot. Outlines Only plots only an outline for the metalization layer. If this check box is not marked the metalization layer is plotted as filled. Scale controls the size of the plot and the Etch Factor corrects for over-etching of the board. The units of the Etch Factor are the units established in the Units section of the Setup menu. A positive number indicates the layout is drawn oversize so over-etching returns objects to the correct dimensions. The Etch Factor algorithm intelligently corrects line, step, bend, tee, cross and gap objects to correct for over-etching in both the X and Y directions.

A filled plot for the elliptic lowpass is given in Figure 6-17. The leader lines extend beyond the cross hairs so a board shear at the cross hairs causes the leader to extend to the edge of the board. The plot is available directly to plotters and laser page printers which support HPGL and also in .HPG and .DXF file formats for interface with other drawing, etching and machining tools.

Scores of filters have been verified from design start to final measurements. Several finished PWB suppliers, file conversion service suppliers and a company which machines PWBs for quick turn around participated in these tests to insure a trouble-free process. A listing of suppliers who participated in these

Figure 6-16 =M/FILTER= Layout Print/Plot Setup window.

Table 6-13 *Final =SuperStar= circuit final in microstrip.*

```
'   FILE: MFILF6C.CKT
'   TYPE: Elliptic -- Lowpass
'     Fc: 3300 MHz
' PROCESS: Microstrip

CIRCUIT
SUB ER=2.55 TAND=.0004 RHO=1                    Whi=14.5375
&  TMet=.71 ROUGH=.06 UNITS=.0254               Wlo=179.206
MLI 1 2 W=WI H=H L=LI                           L1=?89.0708
MST 2 3 O=SY NAR=WI W=Whi H=H                   L2=?43.1034
MLI 3 4 W=Whi H=H L=L1                          L3=?156.106
MTE 4 5 100 WT=Whi WS=Whi H=H                   L4=?195.026
MLI 100 101 W=Whi H=H L=L2                      L5=?236.159
MST 101 102 O=SY NAR=Whi W=Wlo H=H              L6=?71.1138
MLI 102 103 W=Wlo H=H L=L3                      L7=?146.272
MEN 103 0 W=Wlo H=H                             L8=?336.135
MLI 5 6 W=Whi H=H L=L4                          L9=?38.1318
MTE 6 7 110 WT=Whi WS=Whi H=H                   L10=?163.957
MLI 110 111 W=Whi H=H L=L5                      L11=?118.36
MST 111 112 O=SY NAR=Whi W=Wlo H=H              L12=?121.142
MLI 112 113 W=Wlo H=H L=L6                      L13=?164.734
MEN 113 0 W=Wlo H=H                             L14=?135.172
MLI 7 8 W=Whi H=H L=L7                          WOUT=69.0148
MTE 8 9 120 WT=Whi WS=Whi H=H                   LOUT=602.863
MLI 120 121 W=Whi H=H L=L8                      WINDOW
MST 121 122 O=SY NAR=Whi W=Wlo H=H              FILTER(50,50)
MLI 122 123 W=Wlo H=H L=L9                      GPH S21 -5 5
MEN 123 0 W=Wlo H=H                             GPH S21 -100 0
MLI 9 10 W=Whi H=H L=L10                        GPH S11 -40 0
MTE 10 11 130 WT=Whi WS=Whi H=H                 FREQ
MLI 130 131 W=Whi H=H L=L11                     SWP 0 6600 67
MST 131 132 O=SY NAR=Whi W=Wlo H=H              OPT
MLI 132 133 W=Wlo H=H L=L12                     0 3300 S11<-100
MEN 133 0 W=Wlo H=H                             3700 6600 S21<-50
MLI 11 12 W=Whi H=H L=L13                       W21=1000
MST 12 13 O=SY NAR=Whi W=Wlo H=H
MLI 13 14 W=Wlo H=H L=L14
MST 14 15 O=SY NAR=Wlo W=WOUT H=H
MLI 15 16 W=WOUT H=H L=LOUT
DEF2P 1 16 FILTER
EQUATE
H=25
VIAR=20
VIAT=0.71
WI=69.0148
```

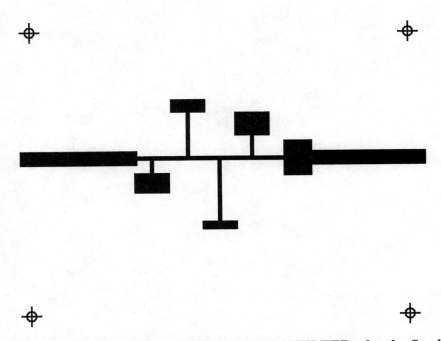

Figure 6-17 *Filled layout created by =M / FILTER= for the final elliptic lowpass.*

tests is given in Appendix A. Also reviewed in Appendix A are common problems encountered in the manufacture of boards and how to avoid these problems.

6.15 References

[1] Eagleware Corporation, 1750 Mountain Glen, Stone Mountain, GA, 30087, USA, TEL (404) 939-0156, FAX (404) 939-0157.
[2] *Technical Overview Manual for IBM PC Software Products*, Eagleware Corporation, Stone Mountain, GA, USA, 1993.
[3] The Effectiveness of Four Direct Search Optimization Algorithms, Randall W. Rhea, IEEE 1987 *MTT-S International Microwave Symposium Digest*, June 9, 1987.

7

Lowpass Structures

This chapter is the first of several which apply the principles covered in earlier chapters to the development of practical filter structures for specific requirements. When necessary, new theories or techniques are introduced which are specific to the structure being considered.

7.1 Overview

The L-C lowpass is a direct application of synthesized prototypes and poses the fewest implementation difficulties of all filter structures. Ideally the same would be true for the distributed lowpass because the synthesis is based on the conversion of L-C filters. Also there is the potential for tighter tolerance on element values. However, difficulties are introduced by the unique characteristics of distributed elements such as reentrance, discontinuities and the realizable range of line impedance.

In this chapter, distributed lowpass filters are studied. The effects of these limitations are considered along with potential methods of mitigating these difficulties.

7.2 Stepped-Impedance All-Pole Lowpass

The stepped-impedance lowpass is a cascade of alternating high and low impedance transmission lines. The high impedance lines act as series inductors and the low impedance lines act as shunt capacitors. This filter structure was used as an example to introduce CAE techniques in Chapter 6. The schematic and a microstrip pictorial are given in Figure 6-2. Chapter 6 contained an example of a lowpass with a 2.25 GHz cutoff and 0.177 dB passband ripple.

In this section a microstrip stepped-impedance 7th order Chebyshev lowpass with a cutoff frequency of 900 MHz and 50 ohm terminations is considered. The nominal substrate dielectric constant is 6.0 with a tolerance of ±0.25 and the board thickness is 25±1 mils. The metalization is 1/2 ounce copper with a nominal thickness of 0.71 mils. The loss tangent and surface roughness are assumed to be 0.001 and 0.01 mils, respectively.

To minimize the effects of line width etching tolerance, a relatively wide minimum line width is selected. To minimize the effects of the step discontinuities and the accuracy of their models, a relatively narrow maximum line width is selected. The initial minimum and maximum line impedances are 16 and 70 corresponding to approximately 18 and 189 mils, respectively.

The initial =M/FILTER= synthesis screen is given in Figure 7-1. The impedance extremes are indeed modest. The high impedance lines are roughly one-third the width of the 50 ohm leaders and the wide lines are less than four times the width of the leaders. The cross hairs are set at 4.1 inches wide by 2.1 inches high. The total length of the filter with 100 mils long leaders exceeds 4.1 inches. The left margin in the Cross hair Setup window of the =M/FILTER= Setup menu is set to −25 mils so the left edge of the left leader extends beyond the cross hairs. The board will be sheered at the cross hairs and mounted in a housing with the center pin of an SMA connector soldered to the leader.

The Write CKT procedure button followed by Write Electrical CKT are selected and =SuperStar= is invoked to display the response of the lowpass modeled with ideal electrical transmission line models. The circuit file written by =M/FILTER= is given in Table 7-1. Actually the file was manually modified in =SuperStar= by replacing the numeric

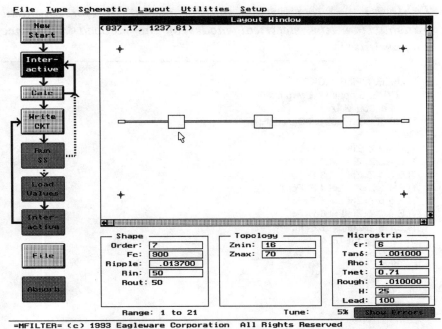

Figure 7-1 *Initial =M/FILTER= screen for the seventh-order stepped-impedance all-pole 900 MHz Chebyshev lowpass.*

value assignments for the line lengths at the output of the filter with L7=L1, L6=L2 and L5=L3. L4 is the length of the series transmission line in the center of the structure. While these manual changes are not mandatory, they maintain symmetry and speed optimization because fewer values must be optimized. The responses of the electrical model of the stepped-impedance lowpass filter before optimization are given as the solid traces in Figure 7-2. The desired cutoff frequency is 900 MHz. Notice the cutoff frequency is approximately 30% too high in frequency and the passband return loss is not equiripple. These are effects of the imperfection of the lumped/distributed equivalences used during synthesis. Also notice the reversal of the attenuation in the upper region of the stopband. This is an effect of reentrance of the transmission line elements.

Table 7-1   =SuperStar= simulator circuit file for an ideal transmission line electrical model of the stepped-impedance lowpass filter.

```
'    FILE: MFILF71.CKT
'    TYPE: Stepped -- Lowpass
'      Fc: 900 MHz
' PROCESS: Electrical Description
CIRCUIT
TLE 1 2 Z=Zhi L=L1 F=Fc
TLE 2 3 Z=Zlo L=L2 F=Fc
TLE 3 4 Z=Zhi L=L3 F=Fc
TLE 4 5 Z=Zlo L=L4 F=Fc
TLE 5 6 Z=Zhi L=L5 F=Fc
TLE 6 7 Z=Zlo L=L6 F=Fc
TLE 7 8 Z=Zhi L=L7 F=Fc
DEF2P 1 8 FILTER
EQUATE
Fc=900
Zhi=70
Zlo=16
L1=?34.4861
L2=?15.3114
L3=?57.2958
L4=?16.9128
L5=L3
L6=L2
L7=L1
WINDOW
FILTER(50,50)
GPH S21 -5 5
GPH S21 -100 0
GPH S11 -30 0
FREQ
SWP 0 3600 73
OPT
0 900 S11<-20
1350 1800 S21<-24
```

The dashed traces in Figure 7-2 are after optimization to recover from the imperfect lumped/distributed equivalences. An OPT section in the WINDOW block is automatically created by the =M/FILTER= program when it writes the circuit file. However, this OPT section is only an estimate. As was pointed out in

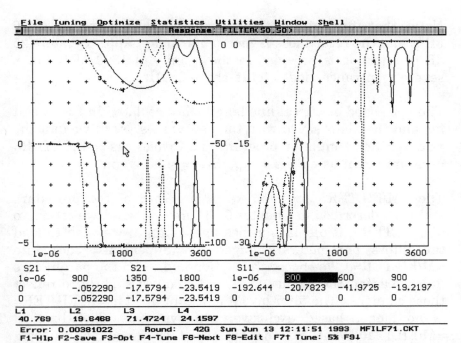

Figure 7-2 *Responses of an ideal transmission line stepped-impedance lowpass using electrical models before (solid) and after optimization (dashed).*

Chapter 6, optimization is not an exact process but a search for a compromise. After observing the resulting responses after a few rounds of optimization, it was noted the passband corner was still too high and the stopband attenuation was lower than desired. The OPT block was modified a few times until the optimized responses were as desired. The final OPT section is given in Table 7-1.

Either the line impedances or the line lengths, or both, could be optimized. As was discussed in Chapter 6, optimization of the line lengths is generally desirable because the tendency in the physical realization is shorter lengths which result in improved stopband performance.

Notice that after optimization the cutoff frequency is close to 900 MHz and the return loss is more equiripple. Approximately 25 dB of rejection is provided in the vicinity of 1800 MHz but a severe reentrance mode occurs above 2 GHz.

The optimized electrical line lengths are replaced in the circuit file and the file is saved with the new values. Next, we quit the =SuperStar= simulator program and invoke the =M/FILTER= program again using the =SuperStar= shell menu.

The =M/FILTER= program keeps a record of the procedure status as described in Chapter 6. Therefore, when we return to =M/FILTER=, it displays a message asking if we wish to load the filters parameters changed in =SuperStar= back into =M/FILTER=. We choose this option and when the parameters finish loading the layout on screen changes slightly to reflect these changes. The filter layout now contained in =M/FILTER= is no longer based exclusively on synthesis algorithms in =M/FILTER=, but also on optimized or tuned values from =SuperStar=. Therefore, many =M/FILTER= functions are unavailable because they would overwrite and destroy the changes supervised by the user in =SuperStar=. Other =M/FILTER= features are still accessible, such as cross hair setup and the ability to write a physical file.

At this point =M/FILTER= contains an electrical process, lowpass filter as modified by =SuperStar=. Next, we consider a microstrip physical implementation of this lowpass. Microstrip is selected in the Process section of the Type menu and then the Write Physical CKT File option of the Write CKT procedure button is selected. =M/FILTER= converts the electrical lines to microstrip, automatically identifies the impedance steps, and absorbs those steps in microstrip by modifying the lengths of the adjacent lines. It then writes a physical model description of the filter for =SuperStar= and invokes a =SuperStar= simulation of the physical filter.

The resulting responses are given in Figure 7-3. The =SuperStar= circuit file is given in Table 7-2. After this file was written by =M/FILTER=, it was again manually modified in =SuperStar= to make L7=L1, L6=L2 and L5=L3 to take advantage of value symmetry.

Notice the responses are nearly identical to the responses of the optimized electrical model of the filter, indicating that the absorption process was successful. Removal of the step discontinuity models from the circuit file but leaving the line lengths at the compensated values increases the cutoff frequency by about 20 MHz, which indicates the steps in this filter have a minimal effect. This is because the frequency of operation is relatively low and the board thickness is only 25 mils. Losses in the microstrip are evidenced in passband attenuation rolloff of

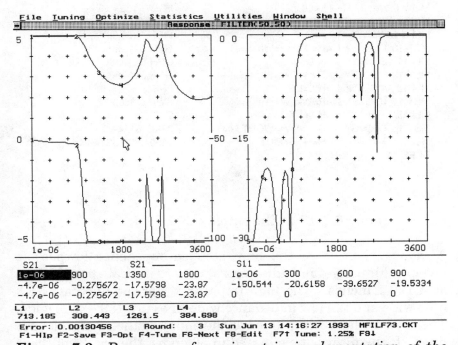

Figure 7-3 Responses of a microstrip implementation of the stepped-impedance lowpass created by =M/FILTER= which converts the electrical design and absorbs the steps.

266 Lowpass Structures

Table 7-2 *Circuit file written by =M/FILTER= for the physical microstrip model of the stepped-impedance lowpass filter which was then manually modified in =SuperStar= to take advantage of value symmetry.*

```
'   FILE: MFILF73.CKT              EQUATE
'   TYPE: Stepped -- Lowpass       H=25
'     Fc: 900 MHz                  WI=36.7163
' PROCESS: Microstrip             LI=100
                                   Whi=18.3144
CIRCUIT                            Wlo=189.43
SUB ER=6 TA=0.001 RHO=1           L1=?713.1854
& TM=0.71 ROUGH=0.01 UNITS=0.0254  L2=?308.4433
MLI 1 2 W=WI H=H L=LI             L3=?1261.497
MST 2 3 O=SY NAR=WI W=Whi H=H     L4=?384.6981
MLI 3 4 W=Whi H=H L=L1            L5=L3
MST 4 5 O=SY NAR=Whi W=Wlo H=H    L6=L2
MLI 5 6 W=Wlo H=H L=L2            L7=L1
MST 6 7 O=SY NAR=Wlo W=Whi H=H    WOUT=36.7163
MLI 7 8 W=Whi H=H L=L3            LOUT=100
MST 8 9 O=SY NAR=Whi W=Wlo H=H    WINDOW
MLI 9 10 W=Wlo H=H L=L4           FILTER(50,50)
MST 10 11 O=SY NAR=Wlo W=Whi H=H  GPH S21 -5 5
MLI 11 12 W=Whi H=H L=L5          GPH S21 -100 0
MST 12 13 O=SY NAR=Whi W=Wlo H=H  GPH S11 -30 0
MLI 13 14 W=Wlo H=H L=L6          FREQ
MST 14 15 O=SY NAR=Wlo W=Whi H=H  SWP 0 3600 73
MLI 15 16 W=Whi H=H L=L7          OPT
MST 16 17 O=SY NAR=Whi W=WOUT H=H 0 900 S11<-20
MLI 17 18 W=WOUT H=H L=LOUT       1350 1800 S21<-24
DEF2P 1 18 FILTER
```

approximately 0.3 dB at the cutoff.

A layout diagram of the final stepped-impedance lowpass is given in Figure 7-4 including cross hairs to indicate board edges at 4.1×2.1 inches. The high impedance narrow lines are 18.3 mils wide and the low impedance wide lines are 189.4 mils wide. The intermediate width lines at the ends are the 50 ohm leader lines. If the filter is terminated in 50 ohms, the length of these leaders affects only the dissipation loss and phase length of the filter.

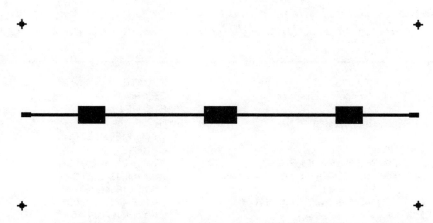

Figure 7-4 *Top view of a final microstrip implementation of the stepped-impedance lowpass.*

7.3 Response Sensitivity to Element Tolerances

The responses given in Figure 7-3 are the best estimate based on the physical microstrip models in the =SuperStar= simulator for the precise dimensions and substrate parameters specified in the circuit file. What are the effects on these responses of dimensional and substrate electrical parameter tolerances? Manufacturing of the substrate and subsequent etching or deposition of the metalization pattern involves tolerances. While tighter tolerances are available at higher cost, typical plastic and PTFE PWB material and manufacturing tolerances are: dielectric constant ±2%, dielectric thickness ±1 mils, and etching tolerance ±1 mil. An etching tolerance is introduced because it is difficult to maintain edge definition during removal of the desired portion of the metalization. Inspection of the stepped-impedance pattern in Figure 7-4 suggests that over-etching would shorten the length and width of the wide lines and would decrease the width and increase the length of the narrow lines.

Next we perform a Monte Carlo test of the stepped-impedance lowpass with pseudo random variations on the dielectric constant, substrate thickness and etching factor. In Table 7-3

Table 7-3 *Circuit file prepared for a Monte Carlo analysis of the substrate thickness, H, the dielectric constant and an etching factor, ETCH.*

```
'   FILE: MFILF73.CKT              EQUATE
'   TYPE: Stepped -- Lowpass       ETCH=?1
'     Fc: 900 MHz                  ETCH=ETCH-1
' PROCESS: Microstrip              H=?25
                                   WI=36.7163+2*ETCH
CIRCUIT                            LI=100+ETCH
SUB ER=?6 TAND=0.001 RHO=1         Whi=18.3144+2*ETCH
& TMet=0.71 ROUGH=0.01 UNITS=0.0254   Wlo=189.43+2*ETCH
MLI 1 2 W=WI H=H L=LI              L1=713.1854-2*ETCH
MST 2 3 O=SY NAR=WI W=Whi H=H      L2=308.4433+2*ETCH
MLI 3 4 W=Whi H=H L=L1            L3=1261.497-2*ETCH
MST 4 5 O=SY NAR=Whi W=Wlo H=H     L4=192.349+ETCH
MLI 5 6 W=Wlo H=H L=L2            WINDOW
MST 6 7 O=SY NAR=Wlo W=Whi H=H     FILTER(50,50)
MLI 7 8 W=Whi H=H L=L3            GPH S21 -5 5
MST 8 9 O=SY NAR=Whi W=Wlo H=H     GPH S21 -100 0
MLI 9 10 W=Wlo H=H L=L4          GPH S11 -30 0
DEF2P 1 10 HALF                   FREQ
HALF 1 2 0                        SWP 0 3600 73
HALF 3 2 0
DEF2P 1 3 FILTER
```

is a circuit file for this filter with the introduction of an etching factor parameter, ETCH. An etching tolerance of ±1 mil is simulated by specifying a ±100% variation on an etch factor of 1 mil, thus producing a range of 0 to 2 mils. After subtracting one, the etch factor becomes ±1 mil. The etch factor, the dielectric constant and the substrate thickness, H, in the circuit file are preceded by a "?," indicating they are to be included in the Monte Carlo analysis. In the Setup Monte Carlo window of the Statistics menu of =SuperStar=we specify a 25 sample run and a seed of zero, and a uniform distribution of the variables with a tolerance of ±100% for the etching factor, ±4% for the substrate thickness and ±4.2% for the dielectric constant. The results of this Monte Carlo analysis are given in Figure 7-5. Variations in the responses are relatively small which suggests that with the specified manufacturing tolerances performance variations should be minimal. However, caution should be exercised in extending the results obtained with this lowpass filter to other filter structures. We will see later that bandpass and bandstop filters are typically much more sensitive to manufacturing variations. In this Monte Carlo example, there was no Yield section in the circuit file so the yield was reported as 100%.

Invoking Sensitivity Analysis from the Statistics menu plots three responses for each variable, one at the minimum variable value, one at the nominal value and one at the maximum value. This quickly reveals that the most significant contributor to the response variations in this example is the etching factor. The minimum line width in this lowpass example is approximately 18.3 mils. If the minimum line width were smaller, the etching factor would represent a greater percentage variation of the line

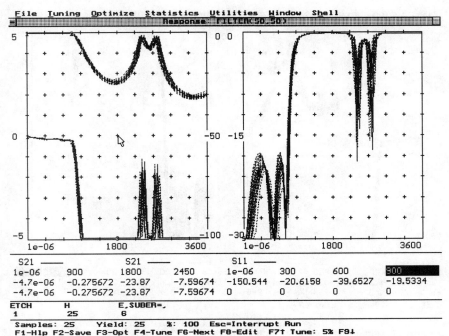

Figure 7-5 *Monte Carlo paint of the microstrip stepped-impedance lowpass with relatively small response variations.*

width. However, the poor stopband rejection due to a low reentrance frequency and the relatively low sensitivity of the responses to the manufacturing tolerances, suggest we should have adopted a more aggressive maximum to minimum line width ratio. Although the tolerance sensitivity would be increased, the lines would be shorter, extending the reentrant frequency and improving the stopband rejection.

Given in Figure 7-6 are the results of a Monte Carlo analysis for a stepped-impedance lowpass filter similar to the previous example except the high impedance line widths are reduced to 7.85 mils and the low impedance line widths are increased to 329 mils. The circuit file is given in Table 7-4. The increased reentrant frequency has improved the stopband rejection over ten decibels around 2 GHz.

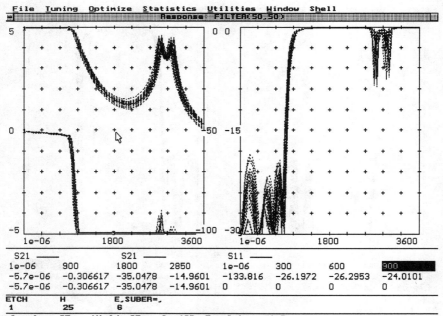

Figure 7-6 *Monte Carlo analysis of the stepped-impedance lowpass with a higher maximum to minimum line width ratio.*

Table 7-4 *=SuperStar= circuit file for stepped-impedance lowpass with higher impedance ratio and ETCH factor added for Monte Carlo analysis.*

```
'   FILE: MFILF73.CKT              EQUATE
'   TYPE: Stepped -- Lowpass       ETCH=?1
'     Fc: 900 MHz                  ETCH=ETCH-1
' PROCESS: Microstrip             H=?25
                                   Wl=36.7163+2*ETCH
CIRCUIT                            Ll=100+ETCH
SUB ER=?6 TAND=0.001 RHO=1         Whi=7.84835+2*ETCH
& TMet=0.71 ROUGH=0.01 UNITS=0.0254  Wlo=329.016+2*ETCH
MLI 1 2 W=Wl H=H L=Ll             L1=569.95-2*ETCH
MST 2 3 O=SY NAR=Wl W=Whi H=H     L2=182.917+2*ETCH
MLI 3 4 W=Whi H=H L=L1            L3=1100-2*ETCH
MST 4 5 O=SY NAR=Whi W=Wlo H=H    L4=102.941+ETCH
MLI 5 6 W=Wlo H=H L=L2            WINDOW
MST 6 7 O=SY NAR=Wlo W=Whi H=H    FILTER(50,50)
MLI 7 8 W=Whi H=H L=L3            GPH S21 -5 5
MST 8 9 O=SY NAR=Whi W=Wlo H=H    GPH S21 -100 0
MLI 9 10 W=Wlo H=H L=L4           GPH S11 -30 0
DEF2P 1 10 HALF                   FREQ
HALF 1 2 0                        SWP 0 3600 73
HALF 3 2 0
DEF2P 1 3 FILTER
```

The narrower lines resulted in greater sensitivity to the etching factor and therefore a larger variation in the responses during the Monte Carlo analysis. It should also be pointed out that a significantly larger ratio of the maximum to minimum line widths, over 40:1, is a more severe test of the accuracy of the microstrip step model and more caution is advised with respect to faith in the simulated results.

7.4 Stepped-Impedance Measured Results

Given in Figure 7-7 is a photograph of an unmounted microstrip prototype of the stepped-impedance lowpass with a high impedance ratio. The measured transmission and return loss responses are given in Figure 7-8. The corner frequency is within the expected range indicated by the Monte Carlo analysis

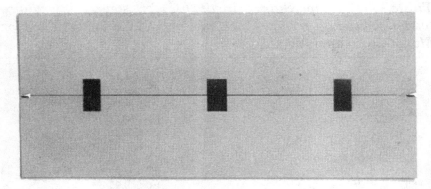

Figure 7-7 *Photograph of microstrip high impedance ratio, 900 MHz, stepped-impedance Chebyshev lowpass on 25 mil thick board with a nominal dielectric constant of 6.0.*

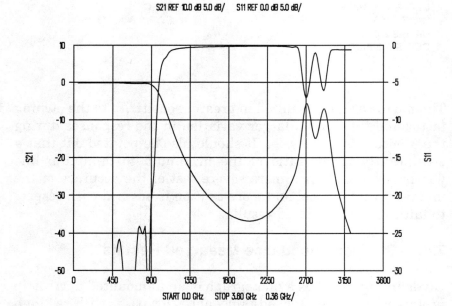

Figure 7-8 *Measured responses of the 900 MHz microstrip stepped-impedance lowpass with a high maximum to minimum line width ratio.*

plotted in Figure 7-6. The exact cutoff is somewhat obscured by the insertion loss rolloff. The measured cutoff based on the return loss is somewhat less than 900 MHz, perhaps 880 MHz which is approximately 2% low. The first reentrance peak as measured occurred at 2690 MHz. The expected reentrant frequency as predicted by the 25 sample Monte Carlo analysis was approximately 2680 to 2740 MHz. The maximum rejection as predicted by Monte Carlo analysis was approximately 35 to 40 dB. The measured value was 37 dB. The predicted return loss varied significantly due to a greater sensitivity. The measured sweep band began at 500 MHz. The two higher frequency return loss maxima agreed very closely in frequency with the predicted nominal values of 580 and 750 MHz and the mean value of these return loss maxima were close to the predicted values of 21 to 24 dB. In general, all parameters are in close agreement with expected values. However, it is important to note that the simulator discontinuity models are not stressed in this example because of the relatively low operating frequency and thin substrate material.

7.5   Stub-Line Lowpass

The previous stepped-impedance lowpass realizes the shunt capacitors of the lowpass prototype as low impedance lines in the transmission path. When the impedance of these lines is made very low, the physical structure more closely resembles stub lines perpendicular to the transmission path. The analogy to symmetrical steps is two opposing stubs connected to the transmission path with a cross model. The analogy to an asymmetric step is a single stub connected in the transmission path with a tee model. Ideally, the responses computed in a simulator program are identical using either step models or stubs connected with tees or crosses regardless of the shape of the structure. In practice, discontinuity model accuracy is a function of dimensional parameters, and when the structure more closely resembles stub-lines, then that form of synthesis and modeling is indicated.

When the lowpass is realized as stepped-impedance cascade and the low impedance line widths are moderate, reentrance has a dominant impact on the stopband attenuation performance. If the shunt capacitors are realized as relatively narrow stub lines, the electrical length of the stubs along the transmission path is short. However, an open stub line which is 30 degrees long at the cutoff frequency has a length of 90 degrees and causes a peak of attenuation at three times the cutoff frequency. The presence of these attenuation peaks causes the stopband characteristics of a stub line lowpass with longer[1] stubs to be substantially different than the stopband of the stepped-impedance lowpass.

Given in Figure 7-9 is an =M/FILTER= screen for a 900 MHz stub-line lowpass with electrical parameters identical to the high impedance ratio stepped-impedance lowpass. The high impedance series lines were set at 70 ohms, identical to the low impedance ratio stepped-impedance lowpass. The low line impedance was set at 24 ohms, significantly lower than with the stepped-impedance lowpass. The nature of the stub-line lowpass is such that a higher line impedance is selected than for the stepped impedance version.

The form selected has a single stub for each shunt capacitor in the lowpass prototype. For a given stub impedance, the double stub form yields shorter stubs and a higher frequency for the attenuation poles. The form which is optimum depends on the specifics of a given application. If the single stub form is suitable for a given application, it is preferred because simulator models are more accurate for the tee than the cross.

The relatively low frequency of this example insures significant spacing between the stub lines. At higher frequencies, an

---

[1]While the stepped-impedance lowpass and stub-line lowpass are fundamentally similar, the term "long" is more natural for stubs in the stub-line lowpass and "wide" is more natural for the low impedance lines in the stepped-impedance structure.

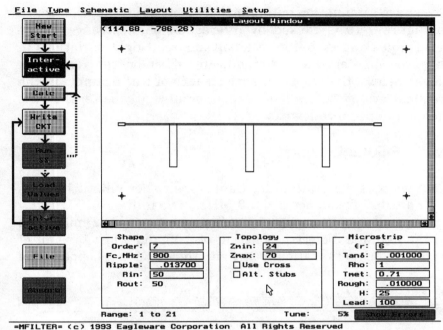

=MFILTER= (c) 1993 Eagleware Corporation All Rights Reserved

Figure 7-9 *900 MHz stub-line lowpass with electrical parameters identical to the stepped-impedance lowpass in Figure 7-7.*

alternating stub form is available which alternates the direction of the stubs to increase the spacing between any two stubs. In general, the stub-line lowpass is physically shorter but wider than the stepped-impedance lowpass.

Given in Figure 7-10 are the responses of the stub line lowpass after optimization in =SuperStar=. The circuit file is given in Table 7-5. Markers two and three are located at attenuation peaks caused by resonance of the stub lines. User selection of the stub line impedance directly affects the electrical length of the stubs and the frequency of the attenuation peaks. Lower line impedance results in shorter stub length and an increase in the attenuation peak frequency. Using double stubs with cross discontinuities approximately halves the stub lengths and doubles the attenuation peak frequencies.

The sharp attenuation minimum between the attenuation peaks in Figure 7-10 is caused by interaction of reentrance in the series lines and stub line resonant modes. Slight variations in these parameters have a significant affect on the responses. Nevertheless, the stopband performance of this example in the frequency range below 3000 MHz is significantly better than the stepped-impedance lowpass.

7.6   Elliptic Lowpass

Next we consider a 5th order Cauer-Chebyshev elliptic lowpass with a cutoff frequency of 1100 MHz. The initial =M/FILTER= screen is shown in Figure 7-11. The substrate is 22 mils thick with a nominal dielectric constant of 2.55 and a loss tangent of 0.0004. The metalization is 2.42 mil thick, 2 ounce copper. The

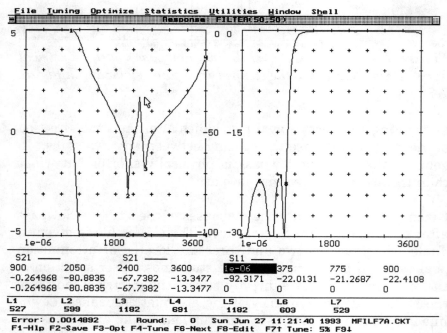

Figure 7-10   *Responses of a 900 MHz microstrip stub-line lowpass with electrical parameters identical to the stepped-impedance lowpass.*

Lowpass Structures 277

Table 7-5 *Circuit file for the 900 MHz microstrip stub line lowpass.*

```
'   FILE: TESTP.CKT                              EQUATE
'   TYPE: Stub -- Lowpass                        H=25
'     Fc: 900 MHz                                WI=36.7163
' PROCESS: Microstrip                            LI=100
                                                 L1=?527
CIRCUIT                                          W1=18.3144
SUB ER=6 TAND=0.001 RHO=1 TMet=0.71             L2=?599
ROUGH=0.01 UNITS=0.0254                          W2=113.147
MLI 1 2 W=WI H=H L=LI                            L3=?1182
MST 2 3 O=SY NAR=WI W=W1 H=H                     W3=18.3144
MLI 3 4 W=W1 H=H L=L1                            L4=?691
MTE 4 5 100 WT=W1 WS=W2 H=H                      W4=113.147
MLI 100 101 W=W2 H=H L=L2                        L5=?1182
MEN 101 0 W=W2 H=H                               W5=18.3144
MLI 5 6 W=W3 H=H L=L3                            L6=?603
MTE 6 7 110 WT=W3 WS=W4 H=H                      W6=113.147
MLI 110 111 W=W4 H=H L=L4                        L7=?529
MEN 111 0 W=W4 H=H                               W7=18.3144
MLI 7 8 W=W5 H=H L=L5                            WOUT=36.7163
MTE 8 9 120 WT=W5 WS=W6 H=H                      LOUT=100
MLI 120 121 W=W6 H=H L=L6                        WINDOW
MEN 121 0 W=W6 H=H                               FILTER(50,50)
MLI 9 10 W=W7 H=H L=L7                           GPH S21 -5 5
MST 10 11 O=SY NAR=W7 W=WOUT H=H                 GPH S21 -100 0
MLI 11 12 W=WOUT H=H L=LOUT                      GPH S11 -30 0
DEF2P 1 12 FILTER                               FREQ
                                                 SWP 0 3600 145
```

minimum and maximum line impedances are moderate, 25 and 95 ohms, with corresponding line widths of approximately 17 mils (Whi) and 156 mils (Wlo). The line lengths are initial synthesized values before optimization in =SuperStar=. =M/FILTER= has modified the lengths to absorb the tee and step discontinuities. WI and LI are the width and length of the 50 ohm input leader and WOUT and LOUT are for the output leader. L1, L4 and L7 are the lengths of the high impedance series transmission lines which replace the series inductors in the L-C prototype. L2 and L5 are the lengths of the high impedance lines which form the inductors in the series L-C branches to ground in the elliptic prototype. L3 and L6 are the lengths of the low impedance lines which form the shunt capacitors in the prototype.

Given in Figure 7-12 are the responses of the 1100 MHz cutoff elliptic lowpass as written by =M/FILTER= (solid) and after optimization to improve the return loss and adjust the cutoff frequency (dashed). The lengths L1 through L7 after optimization are given at the bottom of Figure 7-12. Optimization has reduced the stopband A_{min}. If greater stopband attenuation were needed, new optimization goals with either greater passband ripple or less transition region steepness would be indicated. The circuit file with lengths before optimization is given in Table 7-6.

When the cutoff frequency is increased, the line widths do not change significantly but the line lengths decrease. Eventually the stepped-impedance stub lines approach each other and coupling between stubs becomes important. The alternating stub option assists in managing this problem. Also, as the frequency is increased, the low impedance line sections may overlap the filter leader lines or may become excessively close to the high impedance series lines. A lower value for the high line impedance reduces these difficulties but decreases reentrant mode frequencies and therefore degrades the stopband performance. As the frequency is increased, it becomes necessary to decrease the substrate thickness.

7.7   Elliptic Lowpass Measured Responses

A photograph of an unmounted etched prototype of this elliptic lowpass is given in Figure 7-13 and measured responses to 2750 MHz are given in Figure 7-14. Again excellent agreement is achieved between the computed and measured results. The measured cutoff frequency as established by the return loss is precisely 1100 MHz within measurement resolution. Both finite-frequency transmission zeros (attenuation peaks) also occur very near the desired and computed values. This results in a stopband A_{min} of 28 dB, again very near the computed value.

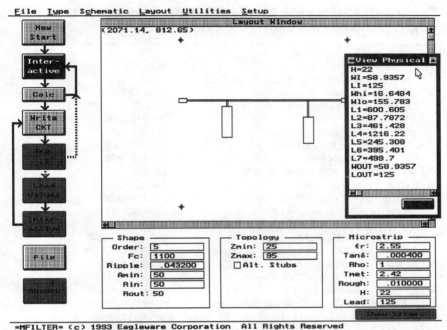

=MFILTER= (c) 1993 Eagleware Corporation All Rights Reserved

Figure 7-11 *Elliptic lowpass initial =M/FILTER= screen with View Physical Values window opened after filter specifications were entered.*

While the two return loss peaks in the computed response do not appear in the measured responses, the general level of the return loss is well over 25 dB as expected. This level of return loss is probably within measurement system capabilities. As with the previous stepped-impedance lowpass, because of the relatively low frequency and thin substrate material, this example is not a severe test of distributed model accuracy.

7.8 Element Collisions

As the cutoff frequency is increased the physical length of transmission line elements decrease. However, the width of the

Table 7-6 *Circuit file for the 5th order microstrip 1100 MHz elliptic lowpass filter.*

```
'    FILE: MFILF78.CKT                    EQUATE
'    TYPE: Elliptic -- Lowpass           H=22
'      Fc: 1100 MHz                      WI=58.9357
' PROCESS: Microstrip                    LI=125
CIRCUIT                                  Whi=16.6484
SUB ER=2.55 TA=0.0004 RHO=1              Wlo=155.783
& TM=2.42 RO=0.01 UNITS=0.0254           L1=?600.605
MLI 1 2 W=WI H=H L=LI                    L2=?87.7872
MST 2 3 O=SY NAR=WI W=Whi H=H            L3=?461.428
MLI 3 4 W=Whi H=H L=L1                   L4=?1216.22
MTE 4 5 100 WT=Whi WS=Whi H=H            L5=?245.308
MLI 100 101 W=Whi H=H L=L2               L6=?395.401
MST 101 102 O=SY NAR=Whi W=Wlo H=H       L7=?499.7
MLI 102 103 W=Wlo H=H L=L3               WOUT=58.9357
MEN 103 0 W=Wlo H=H                      LOUT=125
MLI 5 6 W=Whi H=H L=L4                   WINDOW
MTE 6 7 110 WT=Whi WS=Whi H=H            FILTER(50,50)
MLI 110 111 W=Whi H=H L=L5               GPH S21 -5 5
MST 111 112 O=SY NAR=Whi W=Wlo H=H       GPH S21 -100 0
MLI 112 113 W=Wlo H=H L=L6               GPH S11 -40 0
MEN 113 0 W=Wlo H=H                      FREQ
MLI 7 8 W=Whi H=H L=L7                   SWP 0 4400 221
MST 8 9 O=SY NAR=Whi W=WOUT H=H          OPT
MLI 9 10 W=WOUT H=H L=LOUT               0 1100 S11<-100
DEF2P 1 10 FILTER                        1650 2200 S21<-30
```

lines do not change[1]. This change in the aspect ratio of transmission line elements often results in element overlap (collision).

To illustrate, consider a series of microstrip 6th order elliptic lowpass filters with passband ripple of 0.0432 dB and A_{min} of 42

---

[1]*Unlike stripline, microstrip is dispersive. At low frequencies, the physical width of microstrip for a given impedance is not a function of frequency. At higher frequencies this is not strictly true, but the width variation is small for appropriate substrate thickness. Microstrip dispersion is discussed in Chapter 3.*

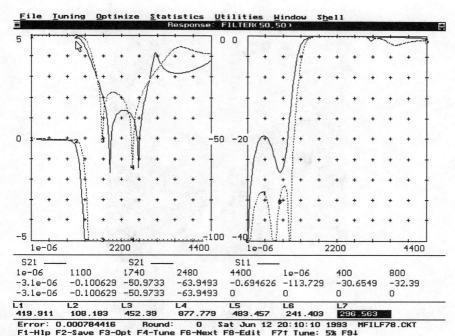

Figure 7-12 *Responses before (solid) and after optimization (dashed) of the 1100 MHz elliptic lowpass microstrip filter.*

dB. The PWB material has a dielectric constant of 2.55 and the metalization is one ounce copper.

The lowpass layout shown in Figure 7-15a is on a 50 mil thick substrate and has a cutoff frequency of 1 GHz. The minimum and maximum line widths are 10 and 250 mils with impedances of 156 and 33 ohms, respectively. The over size including 50 mil long 50 ohm leaders on each end is 2.2 by 1.8 inches.

The filter shown in Figure 7-15b has a cutoff frequency of 2 GHz and uses the same minimum and maximum line widths as the 1 GHz lowpass. The dimensions of the 2 GHz filter including the leaders are 1.1 by 0.9 inches. Each filter in Figure 7-15 is drawn with identical scale. Notice that the wide line segment of the first stub is closer to a corner of the input leader. This has occurred because the line widths in the two filters are equal

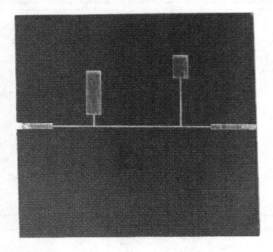

Figure 7-13 *Photograph of the microstrip 1100 MHz elliptic lowpass on 22 mil thick PTFE-glass board.*

but at 2 GHz the line lengths are shorter by a factor of two.

The filter layout in Figure 7-15c has a cutoff frequency of 5.6 GHz. The problem is so severe at this frequency that collisions have occurred. It is obvious that some action is required to resolve this difficulty.

Shown in Figure 7-15d is a 5.6 GHz lowpass on 31 mil thick substrate. The line impedances with 10 and 250 mil wide lines are 135 and 22.7 ohms. These impedances also result in element collision, so the 5.6 GHz lowpass in Figure 7-15d was designed with line impedances of 30 and 90 ohms. The higher minimum impedance narrows the wide segment of the stubs and the lower maximum impedance lengthens the maximum impedance lines, both of which increase element separation. The lower impedance ratio degrades the stopband performance and causes the synthesized response to be less ideal. The later problem is recovered by simulator optimization of line lengths.

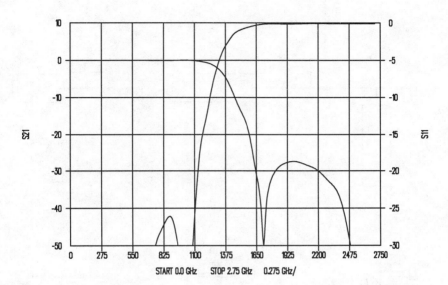

S21 REF 10.0 dB 5.0 dB/   S11 REF 0.0 dB 5.0 dB/

START 0.0 GHz   STOP 2.75 GHz   0.275 GHz/

Figure 7-14 *Measured responses of the 1100 MHz elliptic lowpass on 22 mil thick PTFE-glass board.*

Shown in Figure 7-15e is the 5.6 GHz lowpass on 10 mil thick substrate. The line impedances of 10 and 250 mil wide lines on this substrate are 86 and 8.5 ohms. The layout shown was with line impedances of 15 and 90 ohms. Notice the element clearances are greater and for a given physical separation, the thinner 10 mil thick substrate results in additional inter-element isolation. The overall size of the 5.6 GHz lowpass on 10 mil thick substrate is 0.88 by 0.24 inches.

7.9 References

[1] Eagleware Corporation, 1750 Mountain Glen, Stone Mountain, GA, 30087, USA, TEL (404) 939-0156, FAX (404) 939-0157.

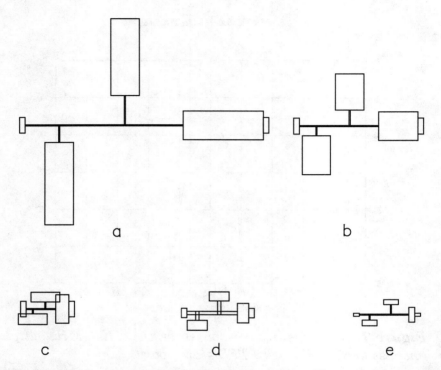

Figure 7-15 *Layouts of elliptic lowpass filters with cutoff frequencies of 1 GHz (a), 2 GHz (b) and 5.6 GHz (c,d,e) and substrate thicknesses of 50 mils (a,b,c), 31 mils (d) and 10 mils (e).*

8

Bandpass Structures

The introduction of a fractional bandwidth parameter for bandpass filters significantly impacts performance and realizability. Over the years, a number of unique distributed bandpass structures have been developed which provide the best possible performance for certain characteristics at the expense of others. There is no one best solution for all applications. The designer who attempts to apply a favorite structure to all problems will not have the success of those who learn to match filter structures and required specifications. Therefore, this chapter is a study of a range of distributed bandpass structures and the advantages and disadvantages of each. We close with a powerful technique for taming the tricky process of tuning bandpass filters of all types.

8.1 Direct-Coupled Bandpass

The direct-coupled bandpass is formed when the lumped/distributed approximate equivalences of Figure 3-15 are applied to the conventional exact-transform lumped element bandpass structure depicted in Figure 4-1. Consider a 3rd order Butterworth lumped element bandpass with a lower 3.01 dB cutoff frequency of 800 MHz and an upper cutoff frequency of 1200 MHz. The schematic of the lumped element network is given in Figure 8-1a. Lumped-element values are found using expressions in Section 4.2 and are listed in column two of Table 8-1. The element values are symmetric, so L1=L3 and C1=C3. The reactances are listed in column three.

The distributed structure in Figure 8-1b is realized by converting the shunt L-C resonators into shorted 90 degree long transmission line stubs. Using the equivalences in Figure 3-15,

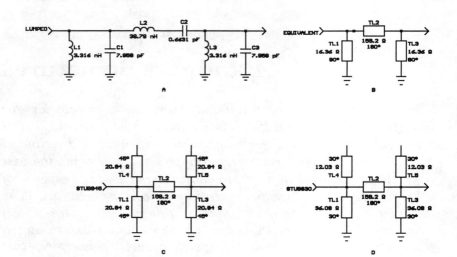

Figure 8-1 *Distributed BP filters formed by replacing lumped with distributed elements. Lumped (a), resonator equivalences (b), 45 degree. lines replacing reactors (c), and 30 degree. lines (d).*

the stub impedance is found to be 16.36 ohms. The lumped series L-C resonator is replaced with a series 180 degree long high-impedance line of 159.2 ohms. These line impedances are listed in column four of Table 8-1.

The responses of the lumped and 90 degree long stub distributed filters are given in Figure 8-2a and 8-2b, respectively. Notice the lumped element response in Figure 8-1a exhibits the typical asymmetry of a conventional lumped bandpass transform as discussed in Chapter 4.

The distributed filter passband center frequency and bandwidth precisely match the lumped element filter. However, notice that the amplitude transmission and group delay responses are symmetrical in the distributed version. With the lumped element bandpass, all transmission zeros occur at dc or infinite

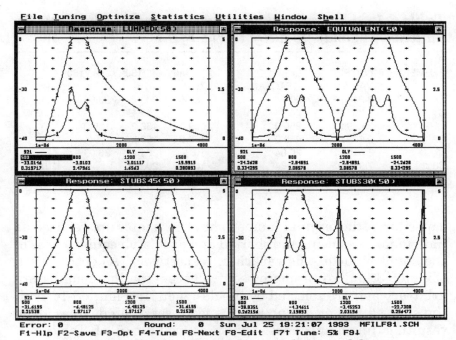

Figure 8-2 Amplitude transmission and group delay responses of the lumped and distributed element filters in Figure 8-1.

frequency. The frequency differentials from the band center to dc and to infinite frequency are quite different, resulting in the asymmetry. However, as predicted by Richard's transform, high side transmission zeros occur at the frequency where the shorted stubs are 180 degree long which is 2 GHz. Therefore, the distributed filter has transmission zeros distributed evenly in an arithmetic sense, resulting in perfect symmetry.

The filter in Figure 8-1b could also be realized with two stubs connected at the same node, thereby forming a cross. Each stub would still be 90 degrees long but the characteristic impedance of each stub would be twice the single stub value. The responses of the single and double stub variations are identical.

The bandpass in Figure 8-1c also utilizes crosses, but the realization procedure is different. In this case, the inductor and

Table 8-1 *Lumped and distributed element values for the Butterworth bandpass filters in Figure 8-1.*

ELEMENT	LUMPED VALUE	LUMPED REACTANCE	90/180° Zo	45/180° Zo	30/180° Z_L	Z_C
L1	3.316 nH	20.84	16.36	20.84	36.09	
C1	7.958 pF	20.84	16.36	20.84		12.03
L2	39.79 nH	250.0	159.2	159.2	159.2	
C2	0.6631 pF	250.0	159.2	159.2		159.2
L3	3.316 nH	20.84	16.36	20.84	36.09	
C3	7.958 pF	20.84	16.36	20.84		12.03

capacitor in the original lumped shunt resonator are realized as individual lines using the shorted and open stub equivalences of the shunt inductor and capacitor, respectively. As described in Chapter 3, unlike resonator equivalences, the line lengths are arbitrary and are selected by the designer. In Figure 8-1c, both the shorted and open line lengths were set at 45 degrees. The resulting line impedances, given in column 5 of Table 8-1, happen to be equal because the tangent of 45 degrees is unity. Notice that transmission zeros occur again at 2 GHz because the capacitive open-stubs behave as shorts at the frequency where they are 90 degrees long. Notice that the bandwidth is too narrow, and therefore the group delay is higher than the lumped bandpass. This is because the 45 degree line length is too long for the lumped/distributed equivalence to hold. This is not the case for 90 and 180 degree lines because they are serving as resonators and not reactors. The bandwidth can be compensated by purposely designing the bandwidth in this case about 15% wider than 400 MHz. This adjusts the line impedances to approximately 30 and 153 ohms for the stub and series lines, respectively.

In the final version of the direct-coupled bandpass, the inductive and capacitive stubs are shortened to 30 degrees. The resulting inductive stub impedance increases to 36.09 ohms and the capacitive stub impedance decreases to 12.03 ohms as indicated in Table 8-1. Notice that the finite transmission zeros have been moved up in frequency to 3 GHz and occur at the reentrant frequency of the series lines resulting in improved stopband performance. Also notice that the group delay has become asymmetric and is more similar to the lumped filter. This simply confirms the indications in Chapter 3 that short lines emulate lumped elements better than longer lines.

It would seem that we have made great progress toward the realization of distributed bandpass filters. However, the careful reader will have noticed that the bandwidth selected for this example is 40%. Just as with the lumped element bandpass, the ratio of element values, in this case the line impedance ratio, increases with decreasing bandwidth. Even though a wide bandwidth has been specified, the ratio of the series and shunt line impedances is 9.7:1 for the filter in Figure 8-1b and 7.6:1 for the filter in Figure 8-1c (5.1:1 after bandwidth correction). Even at 40% bandwidth, the realizable impedance ratios are stressed. If the bandwidth were reduced to 20%, the line impedance ratios are increased by a factor of four making these structures unrealizable. From the equivalent's formula in Figure 3-15, it is easily shown that the ratio of the series to shunt line impedances for the bandpass filter in Figure 8-1b is a factor of two less than the ratio of the series and shunt inductors in the lumped bandpass. For moderate to narrow bandwidth distributed filters, we will need to find alternative structures.

8.2 End-Coupled Bandpass

The end-coupled bandpass is formed with transmission line resonators which are approximately 180 degrees long and are coupled internally and externally with series capacitors formed by gaps between the line ends. The capacitive gaps serve as admittance (J) inverters. Although the lines obviously appear to be series resonators, the inverters present a high-impedance to the resonators which causes them to behave as shunt resonators.

The reactive loading of the series coupling capacitors causes the electrical length of the resonators to be slightly less than 180 degrees and the shortening increases with increasing bandwidth. The following approximate design expressions are from Matthaei, et. al., [1]. The admittances of the resonators are assumed equal to the admittance of the terminations, $Y_o = 1/Z_o$.

$$\frac{J_{01}}{Y_o} = \sqrt{\frac{\pi \; bw}{2g_0 g_1}} \tag{1}$$

$$\frac{J_{n,n+1}}{Y_o} = \frac{\pi \; bw}{2\sqrt{g_n g_{n+1}}} \; , \quad n = 1 \; to \; N-1 \tag{2}$$

$$\frac{J_{N,N+1}}{Y_o} = \sqrt{\frac{\pi \; bw}{2g_N g_{N+1}}} \tag{3}$$

where $J_{n,n+1}$ are the admittance inverter parameters. Then

$$\frac{B_{n,n+1}}{Y_o} = \frac{\dfrac{J_{n,n+1}}{Y_o}}{1 - \left(\dfrac{J_{n,n+1}}{Y_o}\right)^2} \tag{4}$$

$$\theta_n = \pi - 0.5\left[\tan^{-1}\left(\frac{2B_{n-1,n}}{Y_o}\right) + \tan^{-1}\left(\frac{2B_{n,n+1}}{Y_o}\right)\right] \tag{5}$$

where bw is the fractional bandwidth, θ_n is the electrical length of the resonators and $B_{n,n+1}$ is the susceptance of the series coupling capacitors between resonator n and $n+1$.

Gaps in physical transmission lines normally include both series capacitance and shunt capacitance to ground similar to shunt end-effect capacitance. The shunt end-effect capacitance further loads the resonators and decreases the filter center frequency.

One of the primary difficulties with the end-coupled bandpass is the required coupling capacitors become too large to realize as a gap when the bandwidth exceeds a few percent. The external coupling capacitors at the ends of the first and last resonator may exhibit this problem even at narrow bandwidths. These problems worsen with decreasing frequency. It is possible to replace either the end series capacitors, or even all series capacitors, with lumped capacitors. Given in Table 8-2 are the maximum permissible percentage bandwidths for gap spacings of at least 10 mils for a 5th order, 0.0177 dB ripple Chebyshev end-coupled microstrip filter on board material with a relative dielectric constant of 2.2, a substrate thickness (H) of 50 mils and a metalization thickness of 0.71 mils. The Eagleware =SuperStar= and =M/FILTER= gap models based on Kirschning, et. al., [2] are assumed. Other parameters such as the transfer function approximation class, the filter order and board properties such as dielectric thickness and metalization thickness will affect the gap width and therefore the allowable bandwidth. Notice that the bandwidth limitation improves significantly with increasing frequency. Other filter types considered later work well for wider bandwidths, but develop awkward aspect ratios at higher frequencies. Another difficulty with the end-coupled bandpass is extreme length as the frequency is decreased. This and the gap problem suggest reserving the end-coupled structure for higher-frequency applications.

8.3 End-Coupled Bandpass Example

Given in Figure 8-3 is the main =M/FILTER= screen for a 10.25 to 10.45 GHz end-coupled bandpass on 31 mil thick PTFE woven glass with a relative dielectric constant of 2.55. Most of the input parameters are self-explanatory. The minimum gap

Table 8-2 *Maximum bandwidth(%) to avoid microstrip gaps<10 mils for a 5th order, 0.0177 dB ripple end-coupled bandpass. H=50 mils, t_{met}=0.71 mils and ε_r =2.2.*

FREQUENCY	LUMPED END CAPS MAXIMUM BW%	NO LUMPED CAPS MAXIMUM BW%
1000	1.8%	0.04%
2000	3.6%	0.15%
4000	7.2%	0.60%
8000	14.0%	2.3%
12000	20.4%	5.0%

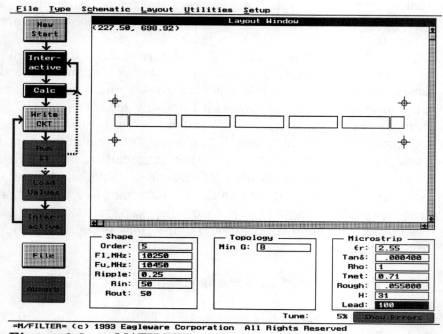

Figure 8-3 *=M/FILTER= screen showing parameters and layout for a 5 section 10 GHz end-coupled microstrip bandpass on 31 mil thick PTFE woven glass material.*

parameter is the gap width below which the gap is assumed unrealizable and =M/FILTER= layout algorithms switch to a gap width to match a specified value for the pad separation for the placement of a lumped capacitor.

The =SuperStar= circuit file written by =M/FILTER= is given in Table 8-2. To simplify later optimization, the circuit file was

Table 8-3 *=SuperStar= circuit file for the 10.35 GHz end-coupled bandpass filter.*

```
'   FILE: MFILF83.CKT                          G3=G2
'   TYPE: End Coupled -- Bandpass              L4=L2
'   Fl: 10250 MHz                              G4=G1
'   Fu: 10450 MHz                              L5=L1
' PROCESS: Microstrip                          G5=G0
                                               WOUT=86.7721
CIRCUIT                                         LOUT=100
SUB ER=2.55 TAND=0.0004 RHO=1                  WINDOW DESIRED
&TMet=0.71 ROUGH=0.055 UNITS=0.0254            FILTER(50,50)
MLI 1 2 W=WI H=H L=LI                          GPH S21 -60 0
MGA 2 3 W=WI G=G0 H=H                          GPH DLY 0 25
MLI 3 4 W=WI H=H L=L1                          FREQ
MGA 4 5 W=WI G=G1 H=H                          SWP 10100 10600 101
MLI 5 6 W=WI H=H L=L2                          OPT
MGA 6 7 W=WI G=G2 H=H                          10250 10450 S11<-20
MLI 7 8 W=WI H=H L=L3                          10100 10150 S21<-43
MGA 8 9 W=WI G=G3 H=H                          10550 10600 S21<-43
MLI 9 10 W=WI H=H L=L4                         WINDOW REENTRANT
MGA 10 11 W=WI G=G4 H=H                        FILTER(50,50)
MLI 11 12 W=WI H=H L=L5                        GPH S21 -60 0
MGA 12 13 W=WI G=G5 H=H                        GPH DLY 0 25
MLI 13 14 W=WOUT H=H L=LOUT                    FREQ
DEF2P 1 14 FILTER                              SWP 19950 20950 101
EQUATE
H=31
WI=86.7721
LI=100
G0=?8.11328
L1=?351.08
G1=?39.9004
L2=?359.165
G2=?43.7451
L3=?359.359
```

manually modified to equate the line lengths and gaps on the output side of the structure to those values at the input side.

Given in Figure 8-4 are the amplitude transmission and group delay responses, for the desired passband on the left and for the first reentrant frequency band on the right. Reentrance occurs just below two times the desired band. Notice the frequency sweep for the desired band is 500 MHz while the specified sweep range for the reentrant band is 1000 MHz. The solid traces on both sides in Figure 8-4 result from the line lengths and gaps in the circuit file as written by =M/FILTER=. The dashed responses are after optimization in =SuperStar= to recover the synthesis inaccuracies which resulted in a slightly narrow bandwidth. The dimensions after optimization are given at the bottom of the =SuperStar= screen in Figure 8-4.

The insertion loss at mid-band for this 1.9% bandwidth filter is approximately 4 dB. The summation of the reactive g-values for a 5th order 0.25 dB ripple Chebyshev is 7.706. The loaded Q is one over the fractional bandwidth, 0.019, or 52.6. Working backward from equation (5) in Chapter 5, the estimated resonator unloaded Q for this filter is approximately 440. Although not verified by the author, the open end resonators are probably extremely susceptible to radiation. To avoid radiation and a significant increase in passband loss, this structure should be mounted in a channel whose width is less than 0.51 inches, which is below cutoff at the upper end of the passband. A thinner substrate would reduce radiation but at the expense of unloaded Q.

8.4   Coaxial End-Coupled Example

Although mechanical construction poses some difficulties, the end-coupled bandpass is better suited for coaxial elements because the facing ends which form the gap are much like a circular parallel-plate capacitor. The increased capacitance provides for much wider bandwidth. Because coaxial elements are typically constructed with a larger longitudinal cross section,

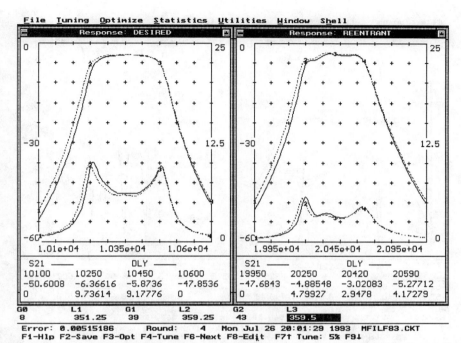

Figure 8-4 *End-coupled bandpass transmission amplitude and group delay responses for the desired band on the left and the first reentrant band on the right. Dashed responses are after optimization.*

the unloaded Q of the elements are naturally higher than with microstrip or stripline.

Given in Figure 8-5 are the amplitude transmission and reflection (return loss) responses of a 6.2 GHz coaxial end-coupled bandpass with a bandwidth of 8.1%. The outer conductor radius is 250 mils and the dielectric is homogeneous solid PTFE with a relative dielectric constant of 2.2. Although the center frequency is 1.6 times lower than the 10.35 GHz microstrip filter and the percentage bandwidth is 4.2 times wider, and both factors would decrease the gap spacing, the coaxial filter end gaps are nearly twice as wide as the microstrip end gaps. Gap and line length dimensions for the coaxial filter are displayed along the bottom of the screen in Figure 8-5.

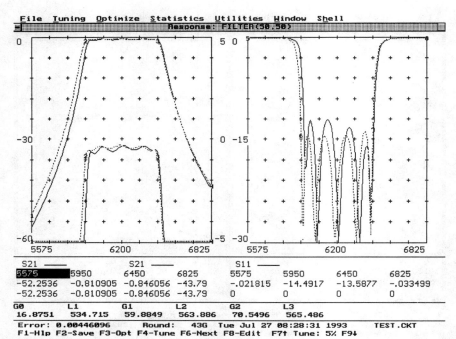

Figure 8-5 *Amplitude transmission and return loss responses of a wider bandwidth coaxial-element end-coupled 6.2 GHz bandpass filter.*

Also notice that the mid-band insertion loss is substantially lower. This is due partly to the higher fractional bandwidth (loaded Q = 12.4) and partly to the higher resonator unloaded Q. Again, working backwards from equation (5) in Chapter 5, the estimated unloaded Q is 1180, about 2.7 times the microstrip resonator unloaded Q.

8.5 Edge-Coupled Bandpass

Significantly greater resonator and external coupling is achieved by arranging half-wave resonators side by side instead of end to end. The total physical length is also reduced. This provides for bandwidths up to about 15%. Additional refinements are discussed later which push the useful bandwidth of the edge-coupled structure to 40% and higher.

Design equations are given by Matthaei, et. al., [3]. The admittance inverter parameters are identical to the end-coupled values are given in equations (1) to (3). The even and odd mode characteristic impedances of each 90 degree long coupled section are then given by

$$Zoe_{n,n+1} = \frac{1}{Y_o}\left[1 + \frac{J_{n,n+1}}{Y_o} + \left(\frac{J_{n,n+1}}{Y_o}\right)^2\right] \tag{6}$$

$$Zoo_{n,n+1} = \frac{1}{Y_o}\left[1 - \frac{J_{n,n+1}}{Y_o} + \left(\frac{J_{n,n+1}}{Y_o}\right)^2\right] \tag{7}$$

Half of each resonator is associated with one of these quarter-wave coupling sections.

The resonators are electrically 180 degrees long at the arithmetic center frequency. End-effect capacitance on each end of the resonator effectively increases the electrical length of the resonators and decreases the filter center frequency. One method of compensation is to shorten the physical length of the resonators at each end so that the ends do not quite reach the centerline of adjacent resonators. A second method is to shorten both lines equally in the quarter-wave sections. The ends and resonator center lines are aligned in this case. Both methods effectively shorten the length of the coupled sections and reduce the coupling. In later examples, using computer simulation we optimize the spacings downward to recover the desired bandwidth and center frequency.

These design expressions result in coupler section line widths which are similar but not exactly equal. This results in width steps at the center of each resonator and in the input and output coupling lines. These may be retained or they may be removed by simply setting all line widths equal and optimizing the section spacings and lengths. This fully recovers the responses and optimization is often required anyway to resolve the end-effect problem described above.

As the bandwidth is increased, the quarter-wave coupling section spacings become small which increases manufacturing difficulties and worsens tolerance effects. As with the end-coupled bandpass, this problem manifests itself in the external coupling sections first. This problem becomes unmanageable above 10 to 15% bandwidth. Three methods are often used to circumvent this difficulty. First, the impedance of the resonators may be increased which increases the required spacings. Second, the first resonator may be externally coupled by tapping instead of using a coupler section. Because the tightest spacings generally occur in the external coupling sections, this significantly extends the useful bandwidth. A third method involves raising the termination resistance presented to the filter by using quarter-wave impedance transformer sections for the input and output leaders. For example, a 50 ohm source driving a quarter-wave 70.7 ohm line section presents a termination resistance of 100 ohms to the filter which has little effect except to increase the required spacing of the external coupling sections. Techniques one and two are supported directly by =M/FILTER=. Technique three is supported by =SuperStar= by tuning or optimization of the spacing after narrowing the input and output leaders to convert them into impedance transformers. These techniques extend the useful edge-coupled filter bandwidth to 40% or more.

8.6   Edge-Coupled Bandpass Example

Given in Figure 8-6 is the layout and input parameters for a narrowband 7th order 5.6 GHz edge-coupled microstrip bandpass on soft ceramic board with a relative dielectric constant of 6.0 and a thickness of 25 mils. The Layout Window cross hairs define a box 2.3 inches wide by 1.0 inches tall. The layout is rotated 21 degrees to align the leaders on a horizontal axis.

The specified corner frequencies are 5525 and 5675 MHz which is a bandwidth of approximately 2.7%. The Slide factor in the Topology box in Figure 8-6 (set at zero for this example) slides the resonators apart by reducing the length of that portion of the resonators which couple to adjacent resonators. A reduction of the coupled line lengths decreases the spacings and is generally undesired in the edge-coupled filter. It is required to realize the geometry of the hairpin bandpass filters considered later, and since the synthesis algorithms are required for the hairpin, the slide factor option is retained for the edge-coupled filter. The substrate parameters are given in the Microstrip box in Figure 8-6.

Given in Figure 8-7 are the amplitude transmission and return loss responses of this 5.6 GHz bandpass after =SuperStar=

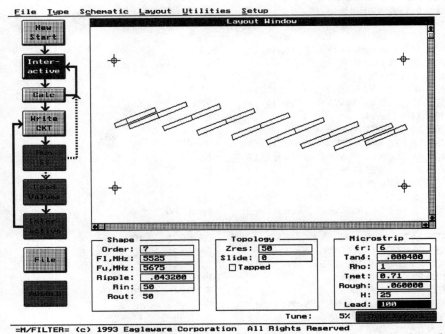

Figure 8-6 *Layout and input parameters for a narrowband edge-coupled 5.6 GHz bandpass on 25 mil thick Rogers Arlon GR6.*

Table 8-4 =SuperStar= circuit file for the narrowband 5.6 GHz edge-coupled bandpass shown in Figure 8-6.

```
'   FILE: MFILF87.CKT                          DEF2P 1 36 FILTER
'   TYPE: Edge Coupled -- Bandpass            EQUATE
'     Fl: 5525 MHz                            H=25
'     Fu: 5675 MHz                            WI=36.7475
' PROCESS: Microstrip                         LI=100
                                              W1=33.22467
CIRCUIT                                       S1=?15.1339
SUB ER=6 TAND=0.0004 RHO=1                    L1=?244.05
& TMet=0.71 ROUGH=0.06 UNITS=0.0254           W2=W1
MLI 1 2 W=WI H=H L=LI                         S2=?75.4645
MST 2 3 O=AS NAR=WI W=W1 H=H                  L2=?243.835
MCP 3 5 6 4 W=W1 S=S1 H=H L=L1                W3=W1
MEN 5 0 W=W1 H=H                              S3=?94.0586
MEN 4 0 W=W1 H=H                              L3=?243.784
MST 6 7 O=AS NAR=W1 W=W2 H=H                  W4=W1
MCP 7 9 10 8 W=W2 S=S2 H=H L=L2               S4=?98.2361
MEN 9 0 W=W2 H=H                              L4=?243.8015
MEN 8 0 W=W2 H=H                              W5=W1
MST 10 11 O=AS NAR=W2 W=W3 H=H                S5=S4
MCP 11 13 14 12 W=W3 S=S3 H=H L=L3            L5=L4
MEN 13 0 W=W3 H=H                             W6=W1
MEN 12 0 W=W3 H=H                             S6=S3
MST 14 15 O=AS NAR=W3 W=W4 H=H                L6=L3
MCP 15 17 18 16 W=W4 S=S4 H=H L=L4            W7=W1
MEN 17 0 W=W4 H=H                             S7=S2
MEN 16 0 W=W4 H=H                             L7=L2
MST 18 19 O=AS NAR=W4 W=W5 H=H                W8=W1
MCP 19 21 22 20 W=W5 S=S5 H=H L=L5            S8=S1
MEN 21 0 W=W5 H=H                             L8=L1
MEN 20 0 W=W5 H=H                             WOUT=WI
MST 22 23 O=AS NAR=W5 W=W6 H=H                LOUT=LI
MCP 23 25 26 24 W=W6 S=S6 H=H L=L6            WINDOW
MEN 25 0 W=W6 H=H                             FILTER(50,50)
MEN 24 0 W=W6 H=H                             GPH S21 -10 10
MST 26 27 O=AS NAR=W6 W=W7 H=H                GPH S21 -100 0
MCP 27 29 30 28 W=W7 S=S7 H=H L=L7            GPH S11 -30 0
MEN 29 0 W=W7 H=H                             FREQ
MEN 28 0 W=W7 H=H                             SWP 5412.5 5787.5 151
MST 30 31 O=AS NAR=W7 W=W8 H=H                OPT
MCP 31 33 34 32 W=W8 S=S8 H=H L=L8            5525 5675   S11<-20
MEN 33 0 W=W8 H=H                             5412.5 5540 S21<-50 W21=100
MEN 32 0 W=W8 H=H                             5750 5787.5 S21<-50 W21=100
MST 34 35 O=AS NAR=W8 W=WOUT H=H
MLI 35 36 W=WOUT H=H L=LOUT
```

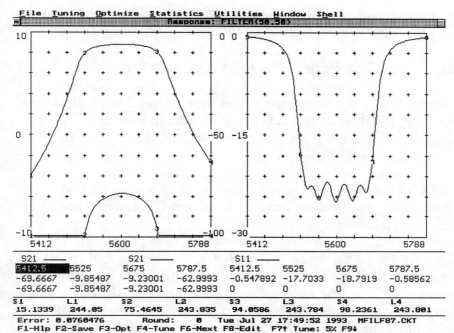

Figure 8-7 *Simulated amplitude transmission and return loss responses of the 5.6 GHz bandpass shown in Figure 8-6.*

optimization of the resonator lengths and spacings to improve the responses resulting from the original =M/FILTER= synthesized values. The =SuperStar= circuit file representing this filter and which lists all pertinent dimensions after optimization is given in Table 8-4. The final bandwidth, as defined by the return loss, is slightly more narrow than 150 MHz. This resulted from optimization targets which specified slightly to much stopband rejection to allow a full 150 MHz passband bandwidth.

Note that the scale of the grid for the passband response in Figure 8-7 is 2 dB per division instead of the normal 1 dB per division used throughout this book. The mid-band insertion loss is just under 6 dB. The narrow bandwidth and thin substrate have conspired to produce significant passband loss, corner rounding and disappearance of the amplitude transmission ripple which is specified as 0.0432 dB.

8.7 5.6 GHz Edge-Coupled Measured Data

Figure 8-8 is a photograph of the narrowband 5.6 GHz edge-coupled PWB with the final design dimensions in Table 8-4. Shown in Figure 8-9 are the measured amplitude transmission and return loss responses for this filter.

SMA connectors were mounted though the housing wall, and the connector round center pins were soldered directly to the filter microstrip leaders.

The measured filter center frequency, bandwidth and insertion loss agreed with the predicted values within the measurement system errors. The return loss was 17 dB, slightly worse than the predicted 22 dB. This is not surprising considering that when measuring a 22 dB return loss, the measured value could be as poor as 17 dB unless the combined return loss of the filter connectors, the transition to microstrip and the measurement system is better than 30 dB.

The responses were measured with the PWB mounted in 0.75 inch wide housing to eliminate the effects of radiation. The measured ultimate stopband rejection is between 70 and 80 dB, suggesting that the narrow housing has discouraged surface waves. Refer to Chapter 5 for additional remarks concerning radiation and surface waves.

8.8 Tapped Edge-Coupled Bandpass

As discussed in Section 8-5, as the bandwidth is extended beyond at 15%, the coupled section spacings become quite small, particularly for the external coupling sections. Shown in Figure 8-10 is the layout and input parameters for an 8 to 12 GHz edge-coupled bandpass with tapped resonators to provide external coupling. The PWB material has a relative dielectric

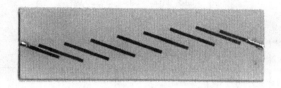

Figure 8-8 *Photograph of the narrowband 5.6 GHz edge-coupled bandpass.*

constant of 6 and a substrate thickness of 25 mils. The metalization is 0.5 ounce copper.

The layout is not the original =M/FILTER= synthesized layout. An initial =M/FILTER= design was created and then the =SuperStar= circuit file was written. Then a number of manual changes were made as described below and =SuperStar= optimization was used to recover the response. The =M/FILTER= layout shown in Figure 8-10 was then created by reading back those changes.

The manual changes were made solely to improve the realizability of the filter. First, the width of the 50 ohm leader lines were decreased from 37 mils to 18 mils to form a 70.7 ohm quarter-wave impedance transformer. This raises the effective termination resistance for the filter from 50 to 100 ohms which has the effect of increasing the required spacings. All resonators were set to equal line widths, a relatively narrow 8 mils (about 95 ohms), again to increase the line spacings. To speed optimization, the line widths, lengths and spacings in the output half of the structure were equated to the input values. The width of the cross hairs in Figure 8-10 is 1.15 inches and the height is 0.6 inches. Even with these modifications to increase the spacings, the spacing between the first and second resonator lines is only 5 mils which is marginal from a realization standpoint.

S21 REF 0.0 dB 10.0 dB/　　S11 REF 0.0 dB 5.0 dB/

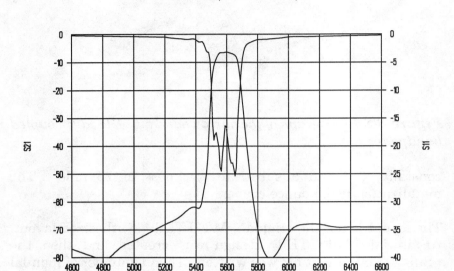

START 4.8 GHz　　STOP 6.80 GHz　0.20 GHz/

Figure 8-9 *Measured amplitude transmission and return loss responses of the narrowband 5.6 GHz edge-coupled bandpass.*

The amplitude transmission and return loss responses after optimization are given in Figure 8-11. Even though the substrate is relatively thin, the insertion loss is under 1 dB because the bandwidth is wide and therefore the loaded Q is low. The wide bandwidth in combination with reentrance results in a maximum rejection of the passband which peaks at approximately 60 dB. Reentrance for half-wave resonators would normally occur at three times the first resonance, or 30 GHz in this case. However, the even and odd mode differential propagation velocity of the coupled sections results in a spurious mode at approximately two times the desired passband. The resulting poor stopband performance of wideband edge-coupled filters is evident in the simulated responses given in Figure 8-12. The homogeneous dielectric media format of stripline avoids even and odd mode differential propagation velocity and should eliminate the spurious responses at two times the passband

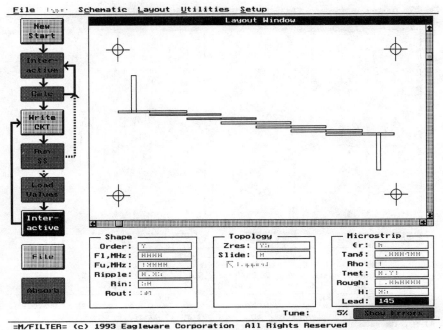

Figure 8-10 *Layout and input parameters for a broadband 8 to 12 GHz edge-coupled bandpass with tapped resonators for external coupling. The layout includes updates created by =SuperStar= optimization.*

frequency. However, even in stripline, slight mis-tuning in the structure can produce spurious responses at twice the passband frequency.

The =SuperStar= circuit file with values saved after manual changes and optimization for this wideband edge-coupled bandpass is given in Table 8-5.

8.9 Hairpin Bandpass

At lower frequencies the aspect ratio of the edge-coupled bandpass is narrow and the length may be excessive. For example, a 9th order 880 MHz edge-coupled bandpass on 50 mil PTFE woven-glass with a relative dielectric constant of 2.2 is

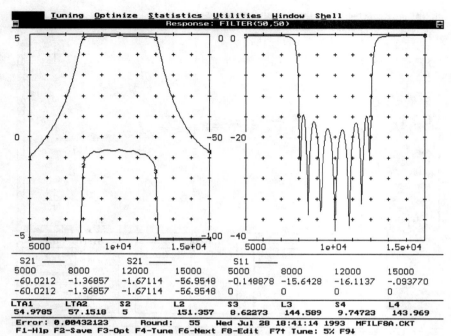

Figure 8-11 *Amplitude transmission and return loss responses of the wideband 8 to 12 GHz edge-coupled bandpass with tapped resonators and quarter-wave impedance-transforming leaders.*

approximately 2.5×24.5 inches. The aspect ratio can be improved and the length reduced by folding the resonators into a "U" shape. This filter structure is often referred to as a hairpin bandpass. The same filter folded into a hairpin with a slide factor of 10 degrees is 2.6×8.0 inches, a substantial reduction in board length and area.

To fold the resonators, it is necessary to introduce a slide factor as described in Section 8-6, otherwise the resonator halves would lie directly adjacent to each other. The introduction of this slide factor reduces the coupled line lengths and therefore reduces the coupling between resonators. This reduces the bandwidth and perturbs the passband ripple. To compensate for sliding, the spacings must be reduced and the resonator line widths adjusted. =M/FILTER= uses an iteration algorithm

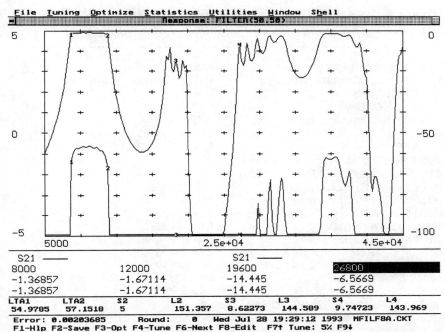

Figure 8-12 *Sweep well into the stopband of the 8 to 12 GHz edge-coupled bandpass.*

which adjusts the width and spacing of a shortened coupling section cascaded with two adjacent lines to match the characteristics of the original quarter-wave section.

The introduction of two bends in each resonator affects the electrical length of the resonators and therefore the center frequency of the filter. =M/FILTER= absorbs the bends using techniques described in Section 3-40.

As the slide factor is reduced the arms of the hairpin resonators become more closely spaced. This introduces resonator self-coupling which narrows the bandwidth and increases the insertion loss of the hairpin filter. Studies of a few examples suggest that resonator self-spacings 2 to 2.5 times larger than the mutual spacings are sufficient. An analysis of self-coupling is provided in Section 8-13.

Table 8-5 =SuperStar= circuit file with the final optimized dimensions for the wideband edge-coupled bandpass.

```
'    FILE: MFILCH8A.CKT                              EQUATE
'    TYPE: Edge Coupled -- Bandpass                  H=25
'     Fl: 8000 MHz                                   WI=18
'     Fu: 12000 MHz                                  Ll=145
' PROCESS: Microstrip                                WTR=8
                                                     LTA1=?54.9785
CIRCUIT                                              LTA2=?57.15179
SUB ER=6 TAND=0.0004 RHO=1                           W2=8
&TMet=0.71 ROUGH=0.06 UNITS=0.0254                   S2=?5
MLI 1 2 W=WI H=H L=Ll                                L2=?151.3566
MTE 4 3 2 WT=WTR WS=WI H=H                           W3=W2
MLI 3 5 W=WTR H=H L=LTA1                             S3=?8.622731
MEN 5 0 W=WTR H=H                                    L3=?144.5887
MLI 4 6 W=WTR H=H L=LTA2                             W4=W2
MST 6 7 O=AS NAR=WTR W=W2 H=H                        S4=?9.747229
MCP 7 9 10 8 W=W2 S=S2 H=H L=L2                      L4=?143.9689
MEN 9 0 W=W2 H=H                                     W5=W2
MEN 8 0 W=W2 H=H                                     S5=S4
MST 10 11 O=AS NAR=W2 W=W3 H=H                       L5=L4
MCP 11 13 14 12 W=W3 S=S3 H=H L=L3                   W6=W2
MEN 13 0 W=W3 H=H                                    S6=S3
MEN 12 0 W=W3 H=H                                    L6=L3
MST 14 15 O=AS NAR=W3 W=W4 H=H                       W7=W2
MCP 15 17 18 16 W=W4 S=S4 H=H L=L4                   S7=S2
MEN 17 0 W=W4 H=H                                    L7=L2
MEN 16 0 W=W4 H=H                                    LTB1=LTA1
MST 18 19 O=AS NAR=W4 W=W5 H=H                       LTB2=LTA2
MCP 19 21 22 20 W=W5 S=S5 H=H L=L5                   WOUT=WI
MEN 21 0 W=W5 H=H                                    LOUT=145
MEN 20 0 W=W5 H=H                                    WINDOW
MST 22 23 O=AS NAR=W5 W=W6 H=H                       FILTER(50,50)
MCP 23 25 26 24 W=W6 S=S6 H=H L=L6                   GPH S21 -5 5
MEN 25 0 W=W6 H=H                                    GPH S21 -100 0
MEN 24 0 W=W6 H=H                                    GPH S11 -40 0
MST 26 27 O=AS NAR=W6 W=W7 H=H                       FREQ
MCP 27 29 30 28 W=W7 S=S7 H=H L=L7                   SWP 5000 15000 161
MEN 29 0 W=W7 H=H                                    OPT
MEN 28 0 W=W7 H=H                                    8000 12000 S11<-30
MST 30 31 O=AS NAR=W7 W=WTR H=H                      5000 6500 S21<-50 W21=1000
MTE 34 33 35 WT=WTR WS=WOUT H=H                      13500 15000 S21<-50 W21=1000
MLI 34 32 W=WTR H=H L=LTB1
MEN 32 0 W=WTR H=H
MLI 33 31 W=WTR H=H L=LTB2
MLI 35 36 W=WOUT H=H L=LOUT
```

While the hairpin significantly reduces the size of lower frequency edge-coupled filters, as the frequency is increased the aspect ratio of each resonator becomes more square and folding is both impractical and saves less space. The frequency at which this is so is a function of the board thickness, the relative dielectric constant and the impedance of the resonators. Other factors being equal, lower frequencies, higher dielectric constants, thinner substrates and higher resonator impedances produce aspect ratios more likely to benefit from folding.

The reduced spacings resulting from the reduction in the length of the coupled sections reduces somewhat the practical bandwidth range of the hairpin. This factor is minimal at low frequencies, but is more significant at higher frequencies where the slide factor must be increased to separate the resonator arms. The same three techniques which are available for extending the edge-coupled bandwidth range are available for the hairpin; tapping the end resonators, increasing the resonator impedance, and using quarter-wave impedance transformers for the input and output leaders.

Radiation is less significant in hairpin filters than in the edge-coupled bandpass. This is discussed in Section 5-8.

8.10 Hairpin 1.27 GHz Example

Shown in Figure 8-13 is the =M/FILTER= layout and input parameter screen for a 5th order 1.27 GHz, 40 MHz wide, hairpin bandpass filter on 25 mil thick board with a relative dielectric constant of 6.0. One ounce copper with a metalization thickness of 0.71 mils was used. At this relatively low center frequency the resonators are long and narrow even with 50 ohm resonator line impedance. This makes folding both desirable and straightforward. The specified slide factor is 12 degrees on each side of the hairpin.

The total length of the hairpin resonators is approximately 180 degrees so the length from the center to either end of the

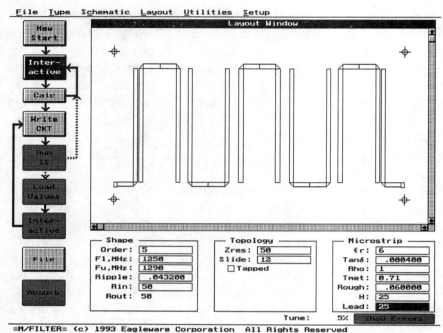

Figure 8-13 *Layout and input parameter screen for a fifth-order 1.27 GHz hairpin bandpass on 25 mil thick board with a relative dielectric constant of 6.0.*

resonator is 90 degrees. Of this 90 degrees, 12 degrees are "slid" out of the coupled section into the uncoupled segment of the resonator. A smaller slide factor tends to significantly shorten the length of the overall filter with only a modest increase in the relative width. The cross hair width and height in Figure 8-13 are 2.45 and 1.25 inches.

The line widths of each of the coupled sections are slightly different, particularly between the external and internal sections. For the sake of elegance, the circuit file written by the synthesis program was manually modified to make all line widths equal (32.44 mils). When the line widths are similar, changing them to equal width is easily recovered by optimization. The file was also manually modified to take advantage of dimensional symmetry about the center of the structure. This reduces the length of the circuit file and reduces the number of variables to optimize. The line lengths and spacings were then optimized in the simulator to fine tune the center frequency, bandwidth and return loss. The circuit file with optimized values is given in Table 8-6.

The simulated amplitude transmission and return loss responses are given in Figure 8-14. The bandwidth as defined by the return loss is just under 40 MHz. The resulting loaded Q, approximately 33, is fairly high for a 1.27 GHz filter on 25 mil substrate. The midband insertion loss predicted by the simulator is 5.7 dB. A thicker substrate such as 50 or 62 mils would reduce the insertion loss. 50 mil board results in approximately 3.5 dB insertion loss.

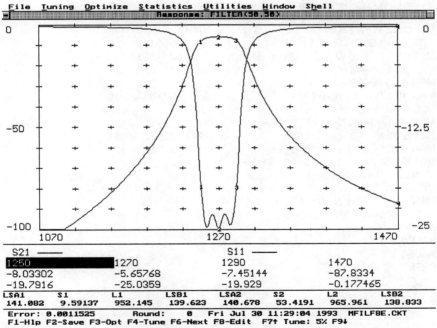

Figure 8-14 *Amplitude transmission and return loss response for the 1.27 GHz hairpin with a bandwidth just under 40 MHz on 25 mil thick board with a relative dielectric constant of 6.0.*

Table 8-6 *Circuit file for the 1.27 GHz hairpin bandpass which has been modified to take advantage of structure symmetry.*

```
'   FILE: MFILF8E.CKT                          EQUATE
'   TYPE: Hairpin -- Bandpass                  H=25
'    Fl: 1250 MHz                              WI=36.7094
'    Fu: 1290 MHz                              LI=25
' PROCESS: Microstrip                          W1=32.44
                                               LSA1=?141.082
CIRCUIT                                        S1=?9.59137
SUB ER=6 TAND=0.0015 RHO=1                     L1=?952.145
&TMet=0.71 ROUGH=0.06 UNITS=0.0254             LSB1=?139.623
MLI 1 2 W=WI H=H L=LI                          W2=W1
MST 2 3 O=AS NAR=WI W=W1 H=H                   LSA2=?140.678
MLI 3 4 W=W1 H=H L=LSA1                        S2=?53.4191
MBN 4 5 O=CH W=W1 H=H                          L2=?965.961
MCP 5 6 7 8 W=W1 S=S1 H=H L=L1                 LSB2=?138.833
MEN 6 0 W=W1 H=H                               W3=W1
MEN 8 0 W=W1 H=H                               LSA3=?138.888
MBN 7 9 O=CH W=W1 H=H                          S3=?66.4369
MLI 9 10 W=W1 H=H L=LSB1                       L3=?965.916
MST 10 11 O=AS NAR=W1 W=W2 H=H                 LSB3=?138.851
MLI 11 12 W=W2 H=H L=LSA2                      WINDOW
MBN 12 13 O=CH W=W2 H=H                        FILTER(50,50)
MCP 13 14 15 16 W=W2 S=S2 H=H L=L2             GPH S21 -100 0
MEN 14 0 W=W2 H=H                              GPH S11 -25 0
MEN 16 0 W=W2 H=H                              FREQ
MBN 15 17 O=CH W=W2 H=H                        SWP 1070 1470 201
MLI 17 18 W=W2 H=H L=LSB2                      OPT
MST 18 19 O=AS NAR=W2 W=W3 H=H                 1250 1290 S11<-30
MLI 19 20 W=W3 H=H L=LSA3                      1220 1235 S21<-50
MBN 20 21 O=CH W=W3 H=H                        1305 1320 S21<-50
MCP 21 22 23 24 W=W3 S=S3 H=H L=L3
MEN 22 0 W=W3 H=H
MEN 24 0 W=W3 H=H
MBN 23 25 O=CH W=W3 H=H
MLI 25 26 W=W3 H=H L=LSB3
DEF2P 1 26 HALF
HALF 1 2 0
HALF 3 2 0
DEF2P 1 3 FILTER
```

8.11 1.27 GHz Hairpin Measured Data

The 1.27 GHz hairpin was constructed by etching 25 mil Arlon GR6 with a relative dielectric constant of 6.0, and an estimated loss tangent of 0.002 and roughness of 0.06 mils. The metalization is 0.5 ounce copper. A photograph of the PWB is given in Figure 8-15. The board was mounted in a 2.54 inch square housing with SMA connectors threaded into the housing wall and the round center pin soldered directly to the PWB.

The measured amplitude transmission and return loss responses are plotted in Figure 8-16. The center frequency is approximately 5 MHz too high which is approximately 0.4%. It is difficult to determine if this is the result of errors in the simulation or the result of tolerance in etching and/or material properties. For example, the tolerance of the dielectric constant of the PWB material is typically no better than ±2% which would result in a center frequency error of ±1% or 12.7 MHz.

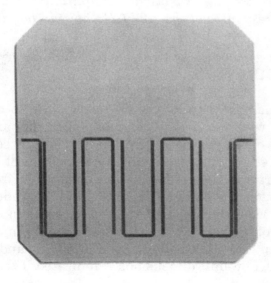

Figure 8-15 *Photograph of the 1.27 GHz hairpin bandpass on Arlon GR6 with a relative dielectric constant of 6.0.*

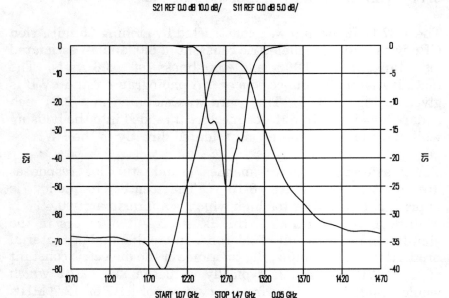

Figure 8-16 *Measured passband amplitude transmission and return loss responses of the 1.27 GHz hairpin bandpass.*

The bandwidth as defined by the return loss is very close to the desired 40 MHz.

The midband insertion loss is also very close to the expected value. The return loss is worse than expected, particularly on the low side of the passband. Even if the center frequency error was due to either the material dielectric constant tolerance or simulator error in the prediction of the resonator frequency, the center frequency would be incorrect but degradation of the return loss would not be expected. The problem could be the return loss of the connector or the transition from coaxial TEM to microstrip. Manual trimming of a prototype and adjustment of the production unit dimensions is indicated if the return loss is to be improved.

8.12 5.6 GHz Hairpin Example

Shown in Figure 8-17 is the layout and input parameter screen for a 7th order 5.6 GHz hairpin filter with a bandwidth of 300 MHz on a 31 mil thick substrate with a relative dielectric constant of 2.55. The resonator line impedance is increased to 84 ohms to improve the folding characteristics. At 50 ohms, the resonator lines are too wide and short to fold well. A slide factor of 20 degrees was chosen which is a compromise between maximum resonator arm spacing to minimize self-coupling and maximum coupling section length for maximum spacings. The external coupling section spacings are 15 mils and the internal spacings range from 53 to 75 mils.

The circuit file with optimized dimensions is given in Table 8-7. Again, the file has been modified to take advantage of structure symmetry. All line widths except the input and output leaders have been set at 33 mils and the spacings and the uncoupled line section lengths optimized for the desired responses. The amplitude transmission and return loss responses after optimization are given in Figure 8-18.

Given in Figure 8-19 are plots of the measured amplitude of the transmission and return loss of a prototype unit constructed using the T-Tech Quick Circuit machine, which directly mills copper from the surface of the PWB. Additional information about this process is given in Section A.2. The PWB was then mounted in a narrow channel to minimize radiation and surface waves and SMA connectors were soldered directly to the PWB.

The passband center frequency is approximately 5.7 GHz, or 1.75% too high. The tolerance of the PWB relative dielectric constant could account for only about ±1% error. The milling tolerances were so small that errors were difficult to detect with a measuring microscope and would account for very little additional error. Milling the PWB does remove approximately 1 mil of dielectric material in the regions where copper is removed. However, this filter was also constructed by etching

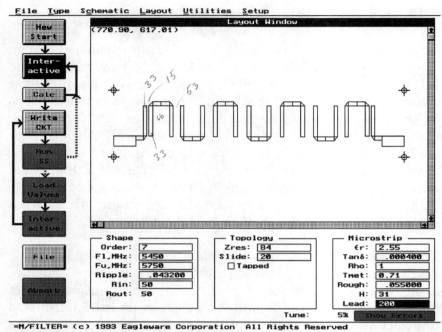

Figure 8-17 *Layout and input parameter screen for a 5.6 GHz hairpin bandpass on 31 mil board with a relative dielectric constant of 2.55.*

PWB material from the same lot and within measurement error, the same results were obtained as with the milled version. One possible conclusion is that the simulator models for microstrip line, coupled line, ends, bends or steps are in error for these board and filter parameters by more than ±0.5% and less than ±1.75%. Tests of other filter structures suggest the problem is more specific and that it is unique to certain hairpin structures. Recall the predicted 1.27 GHz hairpin center frequency was well within the tolerance of the dielectric constant. Whether this is because of different filter specifications or less stressed discontinuity models is not clear.

Also notice the bandwidth is only about 180 MHz instead of the predicted 290 MHz. Again, bandwidth shrinkage was not a problem in the 1.27 GHz hairpin where the bandwidth was

Table 8-7 *Circuit file for the 5.6 GHz hairpin bandpass. The file has been modified to take advantage of structure symmetry.*

```
'    FILE: MFILF8I.CKT                          DEF2P 1 34 HALF
'    TYPE: Hairpin -- Bandpass                  HALF 1 2 0
'     Fl: 5450 MHz                              HALF 3 2 0
'     Fu: 5750 MHz                              DEF2P 1 3 FILTER
' PROCESS: Microstrip                           EQUATE
CIRCUIT                                          H=31
SUB ER=2.55 TAND=0.0004 RHO=1                    WI=85.1107
&TMet=1.42 ROUGH=0.055 UNITS=0.0254             LI=200
MLI 1 2 W=WI H=H L=LI                            W1=33
MST 2 3 O=AS NAR=WI W=W1 H=H                     LSA1=66
MLI 3 4 W=W1 H=H L=LSA1                          S1=15
MBN 4 5 O=CH W=W1 H=H                            L1=285
MCP 5 6 7 8 W=W1 S=S1 H=H L=L1                   LSB1=LSA1
MEN 6 0 W=W1 H=H                                 W2=W1
MEN 8 0 W=W1 H=H                                 LSA2=LSB1
MBN 7 9 O=CH W=W1 H=H                            S2=53
MLI 9 10 W=W1 H=H L=LSB1                         L2=L1
MST 10 11 O=AS NAR=W1 W=W2 H=H                   LSB2=LSB1
MLI 11 12 W=W2 H=H L=LSA2                        W3=W1
MBN 12 13 O=CH W=W2 H=H                          LSA3=LSB2
MCP 13 14 15 16 W=W2 S=S2 H=H L=L2              S3=71
MEN 14 0 W=W2 H=H                                L3=L1
MEN 16 0 W=W2 H=H                                LSB3=LSB1
MBN 15 17 O=CH W=W2 H=H                          W4=W1
MLI 17 18 W=W2 H=H L=LSB2                        LSA4=LSB3
MST 18 19 O=AS NAR=W2 W=W3 H=H                   S4=75
MLI 19 20 W=W3 H=H L=LSA3                        L4=L1
MBN 20 21 O=CH W=W3 H=H                          LSB4=LSB1
MCP 21 22 23 24 W=W3 S=S3 H=H L=L3             WINDOW
MEN 22 0 W=W3 H=H                                FILTER(50,50)
MEN 24 0 W=W3 H=H                                GPH S21 -5 5
MBN 23 25 O=CH W=W3 H=H                          GPH S21 -100 0
MLI 25 26 W=W3 H=H L=LSB3                        GPH S11 -40 0
MST 26 27 O=AS NAR=W3 W=W4 H=H                   FREQ
MLI 27 28 W=W4 H=H L=LSA4                        SWP 5225 5975 76
MBN 28 29 O=CH W=W4 H=H                          OPT
MCP 29 30 31 32 W=W4 S=S4 H=H L=L4            5450 5750 S11<-100
MEN 30 0 W=W4 H=H                                5225 5337.5 S21<-30
MEN 32 0 W=W4 H=H                                5862.5 5975 S21<-30
MBN 31 33 O=CH W=W4 H=H
MLI 33 34 W=W4 H=H L=LSB4
```

predicted well within measurement error. The measured mid-band insertion loss is approximately 6 dB. The predicted insertion loss is just under 3 dB. The increased insertion loss is consistent with the actual bandwidth of the filter.

8.13　Hairpin Resonator Self-Coupling

The above computer simulation modeled the hairpin as coupled line pairs cascaded with uncoupled lines. This model ignores resonator self-coupling. In this section, an analysis of the 5.6 GHz hairpin is conducted which includes resonator self-coupling. This requires a simulator with a multiple-coupled line model. Consider the structure in Figure 8-17. The line on the right side of the first resonator couples not only to the second resonator but also to the line on the left side of the first resonator. This second coupling constitutes the resonator self-coupling.

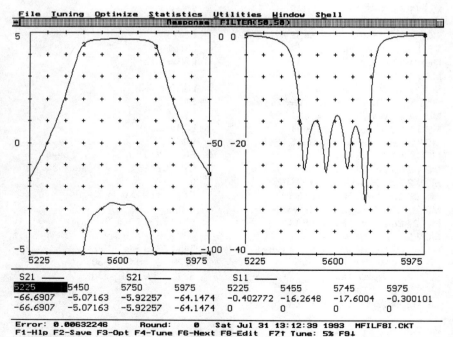

Figure 8-18 *Simulated amplitude transmission and return loss responses of the 5.6 GHz hairpin bandpass.*

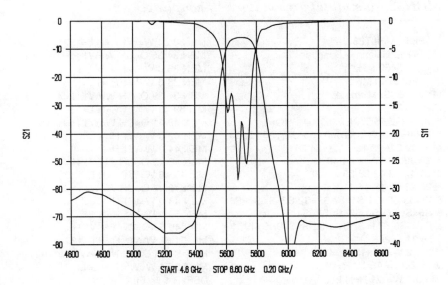

S21 REF 0.0 dB 10.0 dB/   S11 REF 0.0 dB 5.0 dB/

START 4.8 GHz   STOP 6.60 GHz   0.20 GHz/

Figure 8-19 *Measured amplitude transmission and return loss of the 5.6 GHz hairpin bandpass.*

=SuperStar= includes multiple-coupled line models for electrical, microstrip, stripline and slabline. These models consider coupling from each line to adjacent lines on the left and on the right. It does not consider coupling between non-adjacent lines.

To simulate the 5.6 GHz hairpin filter shown in Figure 8-17 with resonator self-coupling, a 16 line (32 node) multiple-coupled microstrip model, MCN32, is included in the circuit file listed in Table 8-8. The uncoupled sections of the resonators are modeled by connecting microstrip lines between the appropriate nodes of the 16 coupled-line model. The bends and microstrip ends in the original hairpin model are also included in this model. A new variable, S_r, is introduced which allows adjustment of the resonator self-spacing which is 132 mils in the constructed prototype. In practice, the resonator self-spacing must equal the length of the connecting uncoupled microstrip

Table 8-8 *5.6 GHz hairpin bandpass =SuperStar= circuit file which includes a 16 line, 32 node, multiple-coupled line model, MCN32, to simulate resonator self-coupling.*

```
'    FILE: MFILT88.CKT              MLI 33 34 W=W1 H=H L=132
'    TYPE: Hairpin -- Bandpass      MBN 34 35 O=CH W=W1 H=H
'     Fl: 5450 MHz                  MEN 36 0 W=W1 H=H
'     Fu: 5750 MHz                  MEN 38 0 W=W1 H=H
' PROCESS: Microstrip               MBN 37 39 O=CH W=W1 H=H
CIRCUIT                             MLI 39 40 W=W1 H=H L=132
SUB ER=2.55 TAND=0.0004 RHO=1       MBN 40 41 O=CH W=W1 H=H
& TMet=1.42 ROUGH=0.055 UNITS=0.0254  MEN 42 0 W=W1 H=H
MCN32 5 8 12 13 17 20 24 25        MEN 44 0 W=W1 H=H
& 29 32 36 37 41 44 48 49          MBN 43 45 O=CH W=W1 H=H
& 50 47 43 42 38 35 31 30          MLI 45 46 W=W1 H=H L=132
& 26 23 19 18 14 11 7 6            MBN 46 47 O=CH W=W1 H=H
& W=33 S0=S1 S1=Sr S2=S2 S3=Sr S4=S3  MEN 48 0 W=W1 H=H
& S5=Sr S6=S4 S7=Sr S8=S4 S9=Sr    MEN 50 0 W=W1 H=H
& S10=S3 S11=Sr S12=S2 S13=Sr S14=S1  MBN 49 51 O=CH W=W1 H=H
& H=31 L=285                        MLI 51 52 W=W1 H=H L=132
MLI 1 2 W=WI H=H L=LI               MST 52 53 O=AS NAR=W1 W=WI H=H
MST 2 3 O=AS NAR=WI W=W1 H=H        MLI 53 54 W=WI H=H L=LI
MLI 3 4 W=W1 H=H L=LSA1             DEF2P 1 54 FILTER
MBN 4 5 O=CH W=W1 H=H               EQUATE
MEN 6 0 W=W1 H=H                    H=31
MEN 8 0 W=W1 H=H                    Sr=?132
MBN 7 9 O=CH W=W1 H=H               WI=85.9356
MLI 9 10 W=W1 H=H L=132             LI=200
MBN 10 11 O=CH W=W1 H=H             W1=33
MEN 12 0 W=W1 H=H                   LSA1=66
MEN 14 0 W=W1 H=H                   S1=15
MBN 13 15 O=CH W=W1 H=H             L1=285
MLI 15 16 W=W1 H=H L=132            S2=53
MBN 16 17 O=CH W=W1 H=H             S3=71
MEN 18 0 W=W1 H=H                   S4=75
MEN 20 0 W=W1 H=H                   LSB8=66
MBN 19 21 O=CH W=W1 H=H             WINDOW
MLI 21 22 W=W1 H=H L=132            FILTER(50,50)
MBN 22 23 O=CH W=W1 H=H             GPH S21 -10 0
MEN 24 0 W=W1 H=H                   GPH S21 -100 0
MEN 26 0 W=W1 H=H                   GPH S11 -30 0
MBN 25 27 O=CH W=W1 H=H             FREQ
MLI 27 28 W=W1 H=H L=132            SWP 5225 5975 76
MBN 28 29 O=CH W=W1 H=H             OPT
MEN 30 0 W=W1 H=H                   5450 5750 S11<-100
MEN 32 0 W=W1 H=H                   5225 5337.5 S21<-30
MBN 31 33 O=CH W=W1 H=H             5862.5 5975 S21<-30
```

lines. To realize the filter, when the resonator self-spacing, S_r, is changed the uncoupled line lengths must also be changed which affects the resonant frequency of the resonators. Mathematically, it is possible to adjust S_r without adjusting the uncoupled line lengths, and that is what we will do in this study. As such, only the $S_r=132$ mil case could be realized.

Listed in Table 8-9 are the resulting upper and lower cutoff frequencies, the bandwidth and insertion loss for values of S_r from 33 to 330. Notice the bandwidth decreases as resonator self-coupling is increased, primarily because the upper cutoff frequency decreases faster than the lower cutoff frequency. Increased resonator self-coupling also decreases the center frequency, decreases the bandwidth and increases insertion loss. However, at $S_r=132$ mils these effects are minimal. The maximum resonator-to-resonator spacing in this filter are 75 mils, so our rule of thumb of self-spacings 2 to 2.5 times the resonator spacings is confirmed in this example. Notice that due to modeling error, the asymptotic value of the insertion loss with large S_r does not quite equal the values obtained when self-coupling is not considered.

8.14 Combline Bandpass

The combline bandpass consists of mutually-coupled resonators which are physically less than a quarter wavelength long and which are grounded at one end and capacitively loaded at the other end. The internal line spacings are sufficiently large that tolerance is not a significant problem, even for moderately wide bandwidths. Resonator length is a design choice, except that it must be less than 90 degrees. At 90 degrees length, the magnetic and electrostatic coupling totally cancel. At less than 90 degrees, magnetic coupling predominates. Resonator line lengths from 45 degrees to less than 30 degrees result in a compact structure with excellent stopband performance. The minimum practical line length is limited by decreased unloaded Q and a requirement for heavy capacitive loading. Lumped capacitors for loading represent a realization nuisance.

Table 8-9 *Cutoff frequencies, bandwidth and insertion loss versus resonator arm spacing, S_r, for the 5.6 GHz hairpin when resonator self-coupling is considered.*

S_r (mils)	F_h (MHz)	F_l (MHz)	BW (MHz)	I.L. (dB)
33	5643	5395	248	5.36
66	5725	5440	285	4.66
99	5742	5445	297	4.48
132	5746	5445	301	4.41
330	5746	5444	302	4.31

However, they also offer a convenient means for tuning. This is particularly critical for narrowband filters where maintaining the required tolerance to avoid tuning is generally impractical.

Combline filters are realized in microstrip and stripline when the convenience of printed filter is important and in machined slabline or rectangular bars[1] when size is less important than high unloaded Q. A combination of fixed chip and mechanical variable capacitors are often used for printed combline filters. At lower frequencies the unloaded Q of capacitors is typically higher than the unloaded Q of inductors and distributed elements. However, at microwave frequencies the unloaded Q of capacitors decreases rapidly with increasing capacitance. This must be carefully considered when selecting loading capacitors for combline filters. Slabline combline filters are often capacitively loaded with threaded screws which penetrate the hollowed rods to form tunable coaxial capacitors.

---

[1]*We will consider rectangular bars to be stripline with a thick strip. Slabline is round rods between flat ground planes. These structures are often realized with an air dielectric and with a ground-plane spacing larger than typical substrate thicknesses. It is this larger ground-plane spacing which results in higher unloaded Q and not the fact they are in air. Except at the higher microwave frequencies, good dielectric materials have less loss than the filter conductors.*

Combline filters have excellent stopband bandwidth because the resonators are electrically short. In addition, the capacitive loading has the effect of doubling the frequency of the first reentrance. In the desired passband, the lines are brought into 90 degree resonance by the loading capacitors. If the lines were 90 degrees long unloaded, the next resonance mode would occur at three times the desired resonant frequency. However, the capacitive reactance is very low at higher frequencies, effectively shorting the loaded end. When shorted at both ends, the lines must be 180 degrees long to resonate. Therefore, the first reentrance mode center frequency is given by

$$f_r \approx \frac{180}{\theta} f_o \tag{8}$$

where θ is the line length in degrees at the desired passband center frequency, f_o. This estimate is accurate for short line lengths and is increasingly conservative for longer θ.

Unlike other distributed structures we will consider, the resonators in combline and interdigital filters couple directly to two other resonators. This poses a difficulty for those simulators which have only two or three-line coupled models. Historically, two methods have been used to overcome this difficulty. Denig [4] describes a simulation method which converts the even and odd mode impedances to values for two coupled lines. Matthaei [5] provides uncoupled wire line models. In practice, both of these techniques require simulation of electrical rather than physical filter descriptions. =SuperStar= provides both electrical and physical multiple-coupled line models, which eliminates the need for structure conversion and allows direct simulation of combline and interdigital structures.

Loss models in circuit simulator programs assume transmission lines are a multiple of a quarter wavelength or are much longer than a wavelength. This is because the models are distributed and the lines are assumed to be longitudinally homogeneous. Transmission line loss is predominantly conductor loss (except at very high frequencies or for low quality dielectrics) when the lines are less than a quarter wavelength. Therefore, the loss predicted by computer simulation will be optimistic for combline filters. A relationship for the unloaded Q of shortened combline resonators is given by Kurzrok [6].

$$Q_e \approx Q_u \sin^2(\theta) \qquad (9)$$

In this author's experience, the above expression is somewhat pessimistic. Nevertheless, as combline resonators are shortened, increased insertion loss is expected.

An often used design technique for combline with external coupling via coupling lines is given in Matthaei, et., al., [7]. The resulting distributed self and mutual line capacitances are then used to determine line widths and spacings for rectangular bars using nomographs by Getsinger [8]. Data and formula for stripline with a thick strip can also serve for rectangular bar design. For slabline, the even and odd mode impedances are also given by Matthaei, et., al., from which the self and mutual capacitances can be derived. The reader should exercise caution when using the slabline impedance formula given in Matthaei which has significant error for larger rod diameters or small spacing. The Stracca, et., al., [9] data for slabline presented in Chapter 3 is more accurate.

Design expressions for tapped combline filters are given by Caspi and Adelman [10]. First, the admittance, Y_a, and center frequency electrical length, θ_o, are chosen for the resonators. The electrical length is selected based on the previous remarks. The resonator admittance is normally selected for minimum loss. For quarter-wavelength resonators, this is approximately $77/\varepsilon_r^{1/2}$. For shorter resonators where conductor loss is increased, a lower impedance is possibly indicated. Then

$$J_{n,n+1} = \frac{bw \; b}{\sqrt{g_n g_{n+1}}} \; , \quad n=1 \, to \, N-1 \tag{10}$$

$$y_{n,n+1} = J_{n,n+1} \; \tan(\theta_o) \tag{11}$$

where

$$b = \frac{Y_a}{2} \left[\frac{\theta_o}{\sin^2\theta_o} + \cot\theta_o \right] \tag{12}$$

$y_{n,n+1}$ are the admittances of the series transmission lines in the equivalent wire-line model of combline. These are used to determine the self capacitance per unit length, C_n/ε, and the mutual capacitance per unit length, $C_{n,n+1}/\varepsilon$ as follows

$$C_1/\epsilon = \frac{376.7}{\sqrt{\epsilon_r}} (Y_a - y_{12}) \tag{13}$$

$$C_n/\epsilon = \frac{376.7}{\sqrt{\epsilon_r}} (Y_a - y_{n-1,n} - y_{n,n-1}) \; , \quad n = 2 \; to \; N-1 \tag{14}$$

$$C_{n,n+1}/\epsilon = \frac{376.7}{\epsilon_r} y_{n,n+1} \; , \quad n = 1 \; to \; N-1 \tag{15}$$

$$C_N/\epsilon = \frac{376.7}{\sqrt{\epsilon_r}} (Y_a - y_{N-1,N}) \tag{16}$$

The loading capacitance for each internal resonator is

$$C_{mid} = \frac{Y_a \cot\theta_o}{\omega_o} \tag{17}$$

For narrowband filters, the loading capacitance values must be precise. The capacitors are then used as tunable elements which compensate for design, mechanical and material tolerances. The loading capacitors for the end resonators are slightly larger than the internal loading capacitors.

The electrical length from the ground end to the tap point on the end resonators, Φ, is given by

$$\Phi = \sin^{-1}\left[\left(\frac{Y_a bw(\cos\theta_o \sin\theta_o + \theta_o)}{2g_o g_1 Y_o} \right)^{1/2} \right] \tag{18}$$

8.15 Coupled Microstrip Combline Example

Given in Figure 8-20 is the layout and input parameter =M/FILTER= screen for a 400 to 440 MHz coupled input and output line 5th order combline bandpass. The PWB material is 1/16 inch thick G-10 (FR-4) with a dielectric constant of 4.8. A low frequency was chosen for this combline example because combline is one of the more compact distributed filter structures which makes it suitable for lower frequency applications. The fact that the dielectric constant of G-10 is higher than PTFE based PWB material also contributes to a smaller size. The fact that the combline structure inherently includes lumped capacitors which may serve as tuning capacitors is also consistent with using G-10 because the relative dielectric constant of this material varies from 4.8 to 5.4 depending on the

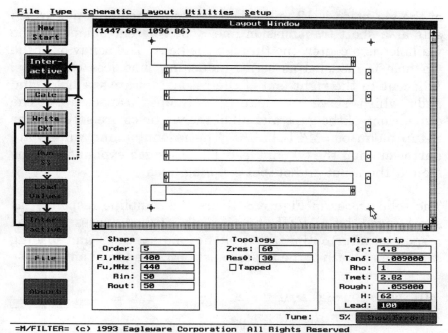

Figure 8-20 *Layout and input parameter screen for a 420 MHz 5th order combline bandpass with input and output coupling lines.*

resin content[1] [11]. Because precision control of the dielectric constant is not necessary for general electronics use, considerable variation can be expected in G-10. The variation in the loss tangent is also large, typically ranging from 0.005 to 0.02 with 0.009 being typical up to 1 GHz.

The cross hair separations in Figure 8-20 are approximately 1.45 inches wide by 1.1 inches high. The objects terminating the resonator lines on the left are via holes to ground. The radius

[1]ε_r=6.1 *for all glass and* ε_r=3.4 *for all resin. The resin content for commercially available G-10 and FR-4 ranges from 25% to 70% which represents a relative dielectric constant range of 5.4 to 4.2. The range for a given manufacturer's process is much smaller but is typically greater than the range in substrates produced for microwave applications.*

of the via holes is 12 mils. The via holes include a pad which acts as a short transmission line. The reference plane for the via holes is a center line through the hole. The behavior of the via hole is modeled as series inductance and loss resistance. Adjacent to the right end of the resonators are via holes and pads which serve to mount the lumped capacitors. The dimensions of the gaps and all of the via holes are setup in the Setup menu of =M/FILTER=. The resonator line widths are narrower than the 50 ohm leader lines which explains the line step at the input and output coupling lines.

The solid traces in Figure 8-21 are the combline responses as synthesized by =M/FILTER=. The dashed traces are the responses after optimization of the line spacings and loading capacitors. Much of the improvement occurred from an increase

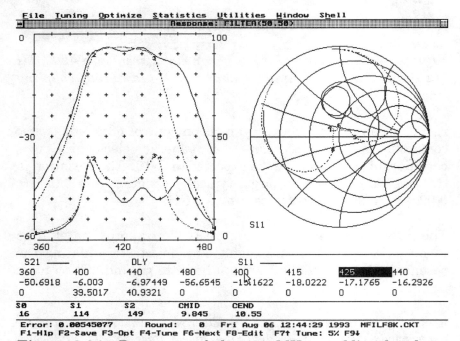

Figure 8-21 *Responses of the 420 MHz combline bandpass before (solid) and after optimization of the spacings and loading capacitance (dashed).*

in the loading capacitor values on the end resonators. The circuit file with optimized element values is given in Table 8-10.

Just as with the edge-coupled and hairpin bandpass filters, the upper bandwidth limit is restricted by intolerably close spacings for the external coupling sections. In the 420 MHz combline, a 60 ohm resonator impedance was selected to increase the spacings. The external spacings for this 9.5% bandwidth case are reasonable at 16 mils. The internal spacings are over 100 mils. Tapped input and output resonators eliminate the external coupling elements and the internal spacings become too small only at a much wider bandwidth.

8.16 Tapped Slabline Combline at 1.27 GHz

This example is a 1270 MHz combline bandpass realized in slabline. Before we consider the details of the slabline filter, we will consider some of the problems of microstrip combline filters which are resolved in the slabline implementation.

When a combline filter is enclosed in a housing, the width of the housing must be sufficiently large to clear not only the resonators but also the space required for the lumped capacitors which load the resonators. For a given physical capacitor size, as the frequency is increased and the resonators become physically shorter, the housing becomes wider than it would need to be which degrades stopband performance. High dielectric constant substrates allow physically small filters and housings which provides a high housing cutoff frequency. Therefore, low dielectric constant substrates are more susceptible to degraded stopband performance, resulting from an increased housing width to accommodate the capacitors. The slabline filter example which follows eliminates this problem by placing the loading capacitors outside the filter cavity.

The only technique available to increase the unloaded Q of transmission line resonators is to increase the transverse size,

Table 8-10 *=SuperStar= circuit file for the 420 MHz microstrip combline bandpass with optimized dimensional values.*

```
'   FILE: MFILF8K.CKT                          EQUATE
'   TYPE: Combline -- Bandpass                 H=62
'     Fl: 400 MHz                              S0=?16
'     Fu: 440 MHz                              S1=?114
' PROCESS: Microstrip                          S2=?149
                                               S3=S2
CIRCUIT                                        S4=S1
SUB ER=4.8 TAND=0.009 RHO=1                    S5=S0
&TMet=2.82 ROUGH=0.055 UNITS=0.0254            W=64
MLI 1 2 W=WI H=H L=LI                          WI=108
MST 2 3 O=AS NAR=WI W=W H=H                    LI=100
MCN14 3 4 5 6 7 8 9 16 15 14 13 12 11 10       WOUT=108
& W=W S1=S0 S2=S1 S3=S2 S4=S3                  LOUT=100
& S5=S4 S6=S5 H=H L=L1                         L1=1233.03
MVH 4 0 R=VIAR H=H T=VIAT                      VIAR=12
MVH 5 0 R=VIAR H=H T=VIAT                      VIAT=1.4
MVH 6 0 R=VIAR H=H T=VIAT                      CMID=?9.845
MVH 7 0 R=VIAR H=H T=VIAT                      CEND=?10.55
MVH 8 0 R=VIAR H=H T=VIAT                      WINDOW
MVH 10 0 R=VIAR H=H T=VIAT                     FILTER(50,50)
MVH 16 0 R=VIAR H=H T=VIAT                     GPH S21 -60 0
MST 9 17 O=AS NAR=W W=WOUT H=H                 GPH DLY 0 100
MLI 17 18 W=WOUT H=H L=LOUT                    SMH S11
CAP 11 19 C=CEND                              FREQ
MVH 19 0 R=VIAR H=H T=VIAT                     SWP 360 480 121
CAP 12 20 C=CMID                              OPT
MVH 20 0 R=VIAR H=H T=VIAT                     400 440 S11<-30
CAP 13 21 C=CMID                              360 372 S21<-50 W21=1000
MVH 21 0 R=VIAR H=H T=VIAT                     468 480 S21<-55 W21=1000
CAP 14 22 C=CMID
MVH 22 0 R=VIAR H=H T=VIAT
CAP 15 23 C=CEND
MVH 23 0 R=VIAR H=H T=VIAT
DEF2P 1 18 FILTER
```

such as the substrate thickness or the coaxial radius[1]. Planar

---

[1]*This strong statement requires qualification. Modern high dielectric constant materials substantially reduce the length of resonators, but contrary to popular belief, they do not increase unloaded Q. Silver metalization and/or exceptionally smooth surfaces can marginally improve unloaded Q, and*

substrates thicker than about 62 mils are expensive and less readily available. Substrates thicknesses to 250 mils are stocked by some manufacturers, but availability of thicker materials is very limited. However, machined slabline or rectangular bar structures are only marginally more expensive for larger arbitrary sizes, especially when the dielectric is air and no dielectric filler is employed. Therein lies the advantage of machined slabline microwave filters; they may be large and therefore have high unloaded Q. This provides for narrower bandwidth filters and lower insertion loss.

Shown in Figure 8-22 is the layout and input parameter screen for an 8th order slabline combline centered at 1270 MHz with a bandwidth of 110 MHz. The line impedance was adjusted so that the resulting rod diameter matched available 209 mil rod stock. To decrease the loss, the rod was silver plated. The resonator length was relatively short (27.1 degrees) to push the reentrance frequency above 8 GHz. No dielectric filler was used and the resulting resonator length is 700 mils. The selected wall to wall spacing (H) was set at 500 mils. While the resistivity of aluminum is higher than copper or silver, the larger size of the housing relative to the rod effectively decreases the contribution of the housing to the overall loss. Therefore, for economy the selected housing material was aluminum. The rod stock was brass because silver plating of aluminum requires plating an additional base metal. The resistivity specified in the substrate section of the input parameter section was one, a rough estimate based on silver rods ($\rho=0.92$) and an aluminum housing ($\rho=1.62$).

The layout in Figure 8-22 does not show via holes or loading capacitor pads because grounding and capacitor construction is mechanical. L1 is the physical distance from the grounded end

superconducting technology has been successfully used to substantially improve unloaded Q. Nevertheless, the most straightforward approach to increasing unloaded Q is increased size.

of the end resonators to the center of the tap point. L2 is the distance from the center of the tap point to the top (loaded) end of the end resonators. The location of this tap point controls the impedance reflected into the filter from the terminations. CMID in Figure 8-22 is the loading capacitance of the six resonators in the middle of the combline. CIN is the capacitance of the input and output resonators. The realization of these capacitors and other construction details are considered in more detail later.

Given in Table 8-11 is the =SuperStar= simulator circuit file written by =M/FILTER=. After a number of optimization rounds, the spacings were fixed on the nearest thousandths of an inch and optimization of the loading capacitors was continued. The file written by =M/FILTER= sets all middle

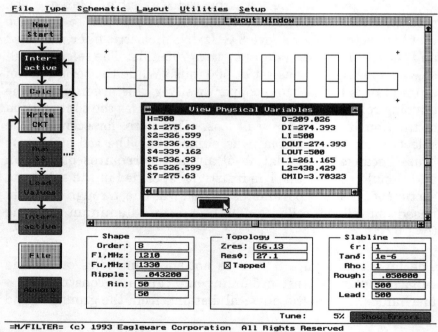

Figure 8-22 =M/FILTER= *layout and input parameter screen for the 1.27 GHz slabline combline bandpass. Overlaid on the screen are the physical dimensions found by =M/FILTER= to match the input specifications.*

Table 8-11   *Circuit file for the 1.27 GHz slabline combline bandpass with optimized values.*

```
'    FILE: TEST.CKT                          S2=331
'    TYPE: Combline -- Bandpass              S3=345
'      Fl: 1210 MHz                          S4=349
'      Fu: 1330 MHz                          S5=S3
' PROCESS: Slabline                          S6=S2
CIRCUIT                                       S7=S1
SUB ER=1 TAND=1e-06 RHO=1                     D=209
&TMet=2.82 ROUGH=0.05 UNITS=0.0254            DI=120
RLI 1 2 D=DI H=H L=LI                         LI=500
RCN16 2 3 4 5 6 7 8 9 0 0 0 0 0 0 0 0         DOUT=DI
& D=D S1=S1 S2=S2 S3=S3 S4=S4 S5=S5           LOUT=LI
& S6=S6 S7=S7 H=H L=L1                        L1=340
RCN16 2 3 4 5 6 7 8 9 17 16 15 14 13 12 11 10 L2=360
& D=D S1=S1 S2=S2 S3=S3 S4=S4 S5=S5           C1=?3.916138
& S6=S6 S7=S7 H=H L=L2                        C2=?3.723059
RLI 9 18 D=DOUT H=H L=LOUT                    C3=?3.69326
CAP 10 0 C=C1                                 C4=?3.68863
CAP 11 0 C=C2                                 WINDOW
CAP 12 0 C=C3                                 FILTER(50,50)
CAP 13 0 C=C4                                 GPH S11 -30 0
CAP 14 0 C=C4                                 GPH S21 -10 0
CAP 15 0 C=C3                                 FREQ
CAP 16 0 C=C2                                 SWP 1170 1370 201
CAP 17 0 C=C1                                 OPT
DEF2P 1 18 FILTER                             1215 1325 S11<-17
EQUATE                                        1170 1190 S21<-34 W21=100
H=500                                         1350 1370 S21<-35 W21=100
S1=269
```

capacitors at equal values. During optimization, each capacitor was allowed an independent value except that symmetry was maintained. The simulated passband responses after optimization of the loading capacitors are given in Figure 8-23. The values in Table 8-11 are after optimization.

8.17   1.27 GHz Combline Measured Data

Given in Figure 8-24 is a photograph of a prototype of this filter. The lower section includes the grounded resonator rods. The

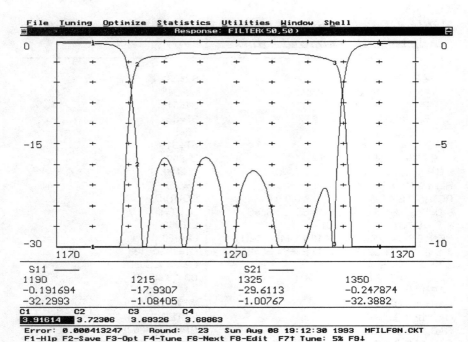

Figure 8-23 *Simulated passband responses of the slabline 1.27 GHz combline bandpass filter after optimization of the spacing and loading capacitors.*

resonators extend above the top of the lower section housing. The upper section includes holes (well-holes) with a diameter of 250 mils which the resonators penetrate for a depth of 300 mils (the resonator rods are 1 inch long). This provides approximately 2.36 pF of capacitance which was found using the =TLINE= [12] program by assuming a coaxial capacitor section. The well-holes are 400 mils deep and there is a small amount of capacitance from the resonator end to the bottom of the well-hole. The remaining loading capacitance is obtained by a brass 2-56 tuning screw threaded into the top housing section which penetrates a 125 mil diameter hole drilled in each resonator rod. Snug-fit PTFE tubing was inserted over the two end tuning screws to insure alignment during filter assembly and to increase the capacitance of the end tuning screws.

Nuts on the tuning screws are used to tighten down the screws. Electrical noise and frequency jumping are eliminated during tuning by slightly tightening these nuts. To facilitate this process, a hollow nut driver with a thumb-wheel is placed over the nut and the screw is tuned through the nut driver with a screwdriver. Dishal's ingenious time-saving tuning procedure is discussed at the end of this chapter.

The slabline resonators are assumed to be between flat ground planes extending infinitely beyond the row of posts. The proximity of the housing end walls lowers the impedance of the end resonators and perturbs the responses. End wall effects are considered by Dishal [13]. The characteristic impedance of a round-rod between ground planes with a third end wall (trough-line) is given approximately by

$$Z_o \approx 138 \log \left[\frac{4h}{\pi d} \tanh \left(\frac{\pi e}{h} \right) \right] \tag{19}$$

where h is the slabline ground-plane spacing, d is the rod diameter and e is the spacing from the rod center to the end wall. This expression is not as accurate as equation 3-51. However, it provides a useful approximation of the effect of an approaching end wall. In the above 1.27 GHz combline, h/d is 2.439 and for large e, Z_o is 67.9 ohms (approximately 1% higher than the more accurate 67.4 ohms from equation 3-51). With $e=h/2$, Z_o is 62.7 ohms. At $e/h=1.22$, Z_o is 67.86, and the end wall effect is nil. Even a rather large end wall effect is readily compensated by increasing the loading capacitance slightly for the end resonators and moving the tap point further from the ground end. For this design, $e/h=1.22$ was selected. In retrospect, for reasons considered in the next paragraph, a closer end-wall spacing would have simplified certain design issues and would have reduced the length of the filter.

The original tap point computation was based on a 50 ohm termination. In order for the 50 ohm connectors to present 50

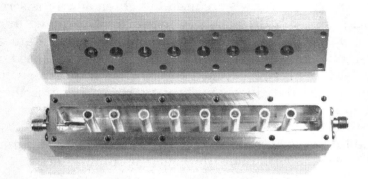

Figure 8-24 *Photograph of a 1.27 GHz 8-section slabline combline. The loading capacitors are formed by resonator penetration of holes in the top section and tuning screw penetration of resonator rod holes.*

ohms to the tap points, the lines extending from the connectors to the resonator tap points must be 50 ohms. Otherwise, they will transform the impedance. From equation (19), the round-rod tap lines from the connectors to the end resonators would need to be approximately 256 mils in diameter. The original tap point from =M/FILTER= was 262 mils above the ground end. It was decided that rods of this diameter would be inconvenient to solder to the end resonators and the connector center pins. Thin flat vanes were selected which were 220 mils wide with the wide dimension along the axis of the resonators. Using a model for stripline in air, the impedance was estimated to be approximately 100 ohms. This was modeled as a round rod line with a diameter of 120 mils. A 100 ohm line which is 500 mils long (19.4 degrees) transforms the 50 ohm termination impedances to approximately $54.5 + j\ 25.6$ ohms. The higher resistance raises the required tap point. The reactance reflected into the end resonators modifies the required loading capacitance. These effects were compensated for by optimizing the tap point location and the loading capacitance of the end resonators. The final tap point was 340 mils from the grounded end of the end resonators.

The length of the tap lines are significantly shortened by placing the connectors on a side wall of the filter at the location of the end resonators. In this example, the distance from the side wall to the resonator tap point is (500-209)/2, or 145.5 mils (5.64 degrees). This is one third the length of the end-mounted connector case and the 50 ohms is transformed thorough a 100 ohm line to $50.36 + j\,7.39$. The resistance is so close to 50 ohms that the design tap point would not need to be modified. The reactance is easily corrected by adjusting the loading capacitance.

The measured responses are given in Figure 8-25. Agreement with the simulated responses is excellent. The measured insertion loss is approximately 0.8 dB. Considering the short length of the resonators, we would expect the predicted loss would be more optimistic than it is. We might have expected the measured loss to be as high as twice the simulated loss or approximately 1.2 dB. This again leads to the conclusion that equation (9) is pessimistic.

8.18 Interdigital Bandpass

As the length of combline resonators approach 90 degrees, the electric and magnetic fields cancel and no coupling occurs. The interdigital structure is formed by grounding the resonators on alternating ends instead of grounding all resonators at adjacent ends as with the combline. Tight coupling occurs at 90 degrees resonator length even for moderately wide spacings. Because the lines are a resonant length, loading capacitance is not required.

The interdigital structure is not as compact as the combline but the resonator unloaded Q is higher. This makes the interdigital particularly well suited when low insertion loss is required. The high unloaded Q is also an advantage for very narrowband filters, except the absence of the loading caps dictates either extremely tight tolerance or slight loading for tuning purposes. Interestingly, the excellent coupling properties make the

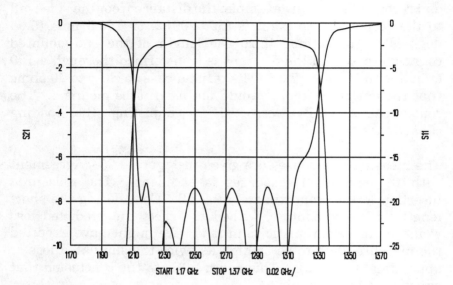

S21 REF 0.0 dB 1.0 dB/   S11 REF 0.0 dB 2.5 dB/

START 1.17 GHz   STOP 1.37 GHz   0.02 GHz/

Figure 8-25 *Measured responses of the 1.27 GHz slabline combline bandpass filter.*

interdigital suitable even for wideband applications, at least up to 70%.

The design procedures for interdigital are similar to combline. Design expressions for the tapped interdigital are also given by Caspi and Adelman [10]. Again, the design begins with the selection of a resonator line admittance, Y_a. Then

$$\theta_1 = \frac{\pi}{2}(1 - \frac{bw}{2}) \tag{20}$$

$$h = \frac{Y_a}{\tan\theta_1} \tag{21}$$

$$J_{n,n+1} = \frac{h}{\sqrt{g_n g_{n+1}}} \ , \quad n = 1 \ to \ N-1 \tag{22}$$

$$y_{n,n+1} = J_{n,n+1} \sin\theta_1 \ , \quad n = 1 \ to \ N-1 \tag{23}$$

$$\Phi = \frac{\sin^{-1}\sqrt{\dfrac{h\sin^2\theta_1}{g_o g_1 Y_A}}}{1 - \dfrac{bw}{2}} \tag{24}$$

θ_1 is the electrical length of the resonators at the lower cutoff frequency. For narrow bandwidth, θ_1 approaches 90 degrees. The mutual admittances, $y_{n,n+1}$, are then used to find the normalized line capacitances using equations (13) through (16) in the previous section for combline. Φ is the electrical length from the end resonator ground to the tap point. The tap process reflects inductance into the end resonators which is compensated for by either lumped capacitance or an extension in length of the end resonators.

Interdigital resonators are 90 degrees long at resonance. Because one end is grounded and one end is open, reentrance occurs at approximately three times the desired passband frequency. One of the drawbacks of the interdigital is that the stopbands are not as wide as combline with short resonators.

8.19 Tapped Interdigital Example

Shown in Figure 8-26 is the layout and input parameter screen for a 5th order tapped interdigital microstrip bandpass centered at 880 MHz. The lower corner is 660 MHz and the upper corner is 1100 MHz for a bandwidth of 50%. This makes construction

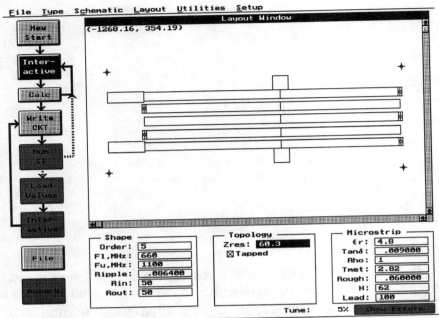

Figure 8-26 *Layout and input parameter screen for the 880 MHz microstrip interdigital bandpass.*

on inexpensive G-10 more feasible since the high loss tangent of this material and the poor tolerance of the dielectric constant are less consequential with wide bandwidth.

The resonator line impedances are specified as 60.3 ohms so they are narrower than the 50 ohm input and output leader lines. Even with 50% bandwidth, the spacings in this filter are 21 and 27 mils which are sufficiently large that manufacturing tolerance should pose no special problems. The transmission line extensions at the open ends of the tapped resonators are much less than 90 degrees long and their effective capacitance compensates the inductance reflected into the end resonators by the tapping process. The compensating line impedance is equal to the specified 60.3 ohm resonator impedance. The mutual coupling between the resonators causes the resonator line widths to be slightly less than the width for uncoupled lines. Therefore the compensating lines are slightly wider than the resonator lines. For elegance, and to remove the effects of the step, the compensating line widths are manually set to equal the resonator line widths in the simulator file and the lengths of the compensating lines are tuned or optimized to adjust for the width change.

The simulator circuit file for the 880 MHz interdigital is given in Table 8-12. The tapped interdigital is modeled as two sets of multiple-coupled lines set end to end at the tap point. L2 is the length of the multiple-coupled line set which is grounded at the end and middle resonators. L1 is the length of the multiple-coupled line set which is terminated with the open compensating lines on the end resonators. Notice the width of the compensating lines, WA, and the width of the resonators, W, have both been set at 60 mils. The resonators are grounded by via holes, in this case with a radius of 12 mils and a metalization thickness of 1.4 mils. The values listed in Table 8-12 are after optimization.

The responses from dc to 3620 MHz computed by the simulator are given in Figure 8-27. Performance is excellent over the frequency range of 880 MHz ±100%. However, the limited stopband performance is evident in the broad sweep. As expected, reentrance occurs at approximately three times the desired passband. If the 320 mil long compensating lines are increased to an electrical length of 30 degrees (approximately 600 mils) so that they are 90 degrees long at the reentrant frequency, then they provide transmission zeros which suppress the first reentrant mode. However, the zeros are not sufficiently broad to suppress the entire stopband from 1760 to 3520 MHz. When these compensating lines are lengthened, the width is reduced so the correct compensating capacitance is presented to the end resonators.

Table 8-12 *Circuit file for the 880 MHz microstrip interdigital filter. The filter has been modified to take advantage of structure symmetry and the values are after optimization.*

```
'   FILE: TEST.CKT                          W=60
'   TYPE: Interdigital -- Bandpass          WI=108
'    Fl: 660 MHz                            LI=20
'    Fu: 1100 MHz                           WOUT=108
' PROCESS: Microstrip                       LOUT=20
CIRCUIT                                      L2=?805
SUB ER=4.8 TAND=0.009 RHO=1                 L1=?970
&TMet=2.82 ROUGH=0.06 UNITS=0.0254          VIAR=12
MLI 1 2 W=WI H=H L=LI                       VIAT=1.4
MCN10 2 3 4 5 6 11 10 9 8 7 W=W S1=S1       WA=60
& S2=S2 S3=S3 S4=S4 H=H L=L2                LA=?320
MCN10 2 3 4 5 6 16 15 14 13 12 W=W          WINDOW
& S1=S1 S2=S2 S3=S3 S4=S4 H=H L=L1          FILTER(50,50)
MLI 6 17 W=WOUT H=H L=LOUT                  GPH S11 -30 0
MVH 7 0 R=VIAR H=H T=VIAT                   GPH S21 -100 0
MVH 9 0 R=VIAR H=H T=VIAT                   MARKER
MVH 11 0 R=VIAR H=H T=VIAT                  180 660 1100 1580
MVH 13 0 R=VIAR H=H T=VIAT                  FREQ
MVH 15 0 R=VIAR H=H T=VIAT                  SWP 0 3520 177
MLI 12 18 W=WA H=H L=LA                     OPT
MEN 18 0 W=WA H=H                           660 1100  S11<-30
MLI 16 19 W=WA H=H L=LA                     0 176     S21<-50 W21=100
MEN 19 0 W=WA H=H                           1584 1760 S21<-50 W21=100
DEF2P 1 17 FILTER
EQUATE
H=62
S1=?21
S2=?27
S3=S2
S4=S1
```

8.20 Coupled Interdigital Example

Shown in Figure 8-28 is the layout and input parameter screen for a three-section coupled input microstrip interdigital 12.45 GHz filter with a bandwidth of 4%. At X-band, 10 mil thick substrate is recommend to minimize discontinuity and radiation issues. However, the unloaded Q of 10 mil board is low for 4% bandwidth. To minimize conductor loss, a 15 mil thick board is

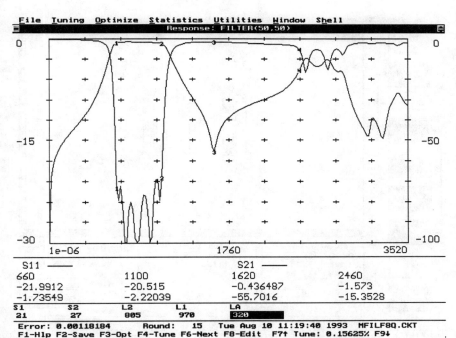

Figure 8-27 *Broad sweep of the 880 MHz microstrip tapped interdigital bandpass showing the desired passband and the first reentrant mode.*

selected instead. The final filter will need to be placed in a covered housing below cutoff to suppress radiation loss.

The circuit file written by =M/FILTER= is given in Table 8-13. The apostrophe at the beginning of certain lines deactivates those lines by turning them into remarks. The lines following remarked out lines modify the filter so the input and output coupling lines are open-circuited instead of shorted with via holes. Also, the file is modified to so the open-circuit input and output coupling lines are fed from the opposite end. The resulting responses are similar to the shorted coupling lines after correction for the removal of via hole discontinuities and the addition of the coupling line ends. The values in the circuit file are after optimization to correct for these modifications. The responses for this filter are given in Figure 8-29.

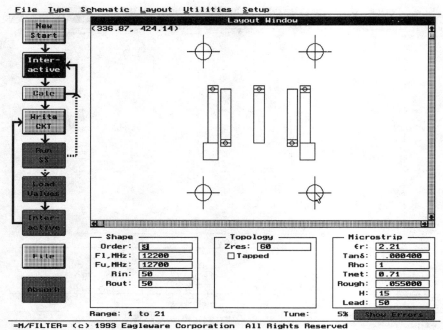

Figure 8-28 *Layout and input parameter screen for a three-section microstrip coupled input interdigital bandpass.*

8.21 Transmission Zeros in Combline and Interdigital

The selectivity of combline and interdigital filters is improved by adding transmission zeros created by coupling resonant lines to the resonators. This works well with either interdigital or combline structures and is illustrated in Figure 8-30 for a tapped slabline combline of 4th order with four all-pole resonators. The zero-forming resonators may be offset as in Figure 8-30a or they may be in-line with the all-pole resonators as shown in Figure 8-20b. The former method allows for and often uses smaller diameter zero-forming resonators. The smaller diameter results in lower unloaded Q for those resonators, but since these resonators are used for transmission zeros, the effect on the passband insertion loss is minimal. The latter method requires that the connectors be placed on the side walls but facilitates computer simulation because the resonators are equal diameter

Table 8-13 *Simulator circuit file for the 12.45 GHz coupled interdigital bandpass. The circuit is modified to use open-circuit input and output coupling lines.*

CIRCUIT	EQUATE
SUB ER=2.21 TAND=0.0004 RHO=1	H=15
&TMet=0.71 ROUGH=0.055 UNITS=0.0254	S0=?13
MLI 1 2 W=WI H=H L=LI	S1=?70
'MST 2 3 O=AS NAR=WI W=W H=H	S2=S1
MST 2 12 O=AS NAR=WI W=W H=H	S3=S0
MCN10 3 4 5 6 7 8 9 10 11 12 W=W	W=31.2054
& S1=S0 S2=S1 S3=S2 S4=S3 H=H L=L1	WI=45.2072
MVH 4 0 R=VIAR H=H T=VIAT	LI=50
MEN 5 0 W=W H=H	WOUT=45.2072
MVH 6 0 R=VIAR H=H T=VIAT	LOUT=50
'MVH 8 0 R=VIAR H=H T=VIAT	L1=?162.1654
MEN 9 0 W=W H=H	VIAR=8
MVH 10 0 R=VIAR H=H T=VIAT	VIAT=0.71
MEN 11 0 W=W H=H	WINDOW
'MVH 12 0 R=VIAR H=H T=VIAT	FILTER(50,50)
MEN 3 0 W=W H=H	GPH S21 -60 0
'MST 7 13 O=AS NAR=W W=WOUT H=H	GPH P21 -180 180
MST 8 13 O=AS NAR=W W=WOUT H=H	SMH S11
MLI 13 14 W=WOUT H=H L=LOUT	FREQ
DEF2P 1 14 FILTER	SWP 11200 13700 101
MLI 1 2 W=WI H=H L=LI	OPT
DEF2P 1 14 FILTER	12300 12600 S11<-20
	11200 11450 S21<-40
	12450 13700 S21<-40

and in-line. In this event, the multiple-coupled slabline model available in =SuperStar= Professional is valid for both the conventional and zero-forming resonator posts. The configurations in Figure 8-30 provide for either one or two zero-forming resonators by placing a zero forming resonator at one or both ends. With two zero-forming resonators, both zeros may be placed below the passband, both may be placed above the passband or one may be placed on each side of the passband, depending on the selectivity and rejection requirements.

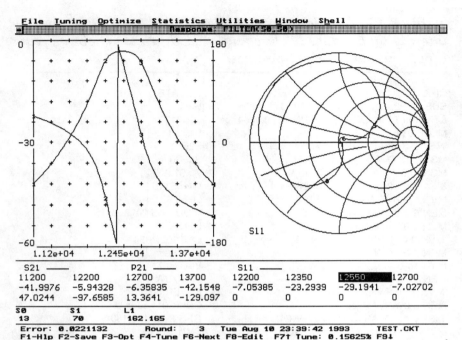

```
File  Tuning  Optimize  Statistics  Utilities  Window  Shell
                        Response  FILTER(50.50)

  S21 ——              P21 ——              S11 ——
11200     12700     12200     13700     12200     12350     12550     12700
-41.9976  -5.94328  -6.35835  -42.1548  -7.05385  -23.2939  -29.1941  -7.02702
47.0244   -97.6585  13.3641   -129.097  0         0         0         0

S0        S1        L1
13        70        162.165
Error: 0.0221132      Round:    3   Tue Aug 10 23:39:42 1993       TEST.CKT
F1-Hlp F2-Save F3-Opt F4-Tune F6-Next F8-Edit  F7↑ Tune: 0.15625% F9↓
```

Figure 8-29 *Amplitude and phase transmission responses (left) and input return loss plotted on a Smith chart (right) for the three-section microstrip interdigital bandpass with open-end coupling lines.*

When external coupling is provided by coupled-line sections as opposed to tapping, both transmission zeros naturally occur on one side of the passband. Whether the zeros occur below or above the passband depend on whether the external coupling lines are open or grounded at the end opposite of the connectors.

The transmission amplitude and group-delay responses of a 7th order 1.27 GHz all-pole combline is given on the left in Figure 8-31. On the right is the same combline with a transmission zero added below and one above the passband. The solid traces are with the zero-forming resonator tuned for notches at approximately 1200 and 1340 MHz. The dashed responses are with the low side notch tuned 20 MHz lower and the high side notch tuned 20 MHz higher.

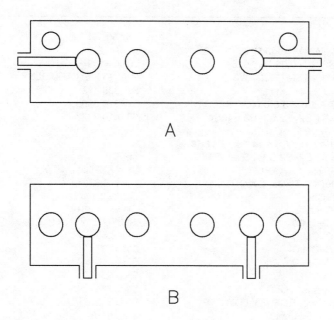

Figure 8-30 *Top-view of tapped slabline combline with four all-pole resonators and two transmission-zeros. (A) uses offset zero-forming resonators and (B) uses all in-line resonators.*

The simulator circuit file for these filters is given in Table 8-14. The combline with transmission zeros is created by modifying the all-pole combline. The in-line resonator structure shown in Figure 8-31b was chosen. First the topology is modified to include the additional resonators but the spacings are set very large so they couple loosely to the tapped resonators. The spacings are then tuned lower and the resonator loading capacitors are adjusted so the zeros become visible in the swept response. The spacings and loading capacitors are adjusted to set the zero frequencies and to adjust the width of the notches.

Table 8-14 *Simulator circuit file for two slabline combline filters, one without transmission zeros (all-pole) and one with two zeros.*

```
CIRCUIT                                      LI=150
SUB ER=1 TAND=1e-06 RHO=1                     DOUT=80
& TMet=2.82 RO=0.06 U=0.0254                  LOUT=150
'ALL_POLE FILTER FOLLOWS                      'ALL_POLE VARIABLES
RLI 1 2 D=DI H=H L=LI                         S1=338.4721
RCN14 2 3 4 5 6 7 8 0 0 0 0 0 0 0 0           S2=396.9503
& D=D S1=S1 S2=S2 S3=S3 S4=S4                 S3=410.0897
& S5=S5 S6=S6 H=H L=L1                        S4=S3
RCN14 2 3 4 5 6 7 8 15 14 13 12 11 10 9       S5=S2
& D=D S1=S1 S2=S2 S3=S3 S4=S4                 S6=S1
& S5=S5 S6=S6 H=H L=L2                        L1=230.7504
RLI 8 16 D=DOUT H=H L=LOUT                    L2=475.6331
CAP 9 0 C=CIN                                 CMID=3.571197
CAP 10 0 C=CMID                               CIN=3.644614
CAP 11 0 C=CMID                               'ELLIPTIC VARIABLES
CAP 12 0 C=CMID                               Sae=240
CAP 13 0 C=CMID                               S1e=307
CAP 14 0 C=CMID                               S2e=410
CAP 15 0 C=CIN                                S3e=421
DEF2P 1 16 ALL_POLE                           S4e=S3e
'FILTER WITH ZEROS FOLLOWS                    S5e=S2e
RLI 1 2 D=DI H=H L=LI                         S6e=S1e
RCN18 20 2 3 4 5 6 7 8 18 0 0 0 0 0 0 0 0 0   Sbe=240
& D=D S1=Sae S2=S1e S3=S2e S4=S3e             L1e=300
& S5=S4e S6=S5e S7=S6e S8=Sbe                 L2e=400
& H=H L=L1e                                   C7e=3.43
RCN18 20 2 3 4 5 6 7 8 18 17 15 14 13 12      C6e=3.645
& 11 10 9 19 D=D S1=Sae S2=S1e S3=S2e         C4e=3.623
& S4=S3e S5=S4e S6=S5e S7=S6e                 C5e=C4e
& S8=Sbe H=H L=L2e                            C3e=C4e
RLI 8 16 D=DOUT H=H L=LOUT                    C2e=3.64
CAP 19 0 C=Cae                                C1e=4.22
CAP  9 0 C=C1e                                Cae=?4.17
CAP 10 0 C=C2e                                Cbe=?3.28
CAP 11 0 C=C3e                                WINDOW
CAP 12 0 C=C4e                                ALL_POLE(50)
CAP 13 0 C=C5e                                GPH S21 -100 0
CAP 14 0 C=C6e                                GPH DLY 0 100
CAP 15 0 C=C7e                                FREQ
CAP 17 0 C=Cbe                                SWP 1120 1420 121
DEF2P 1 16 ELLIPTIC                           WINDOW
EQUATE                                        ELLIPTIC(50)
'COMMON VARIABLES                             GPH S21 -100 0
```

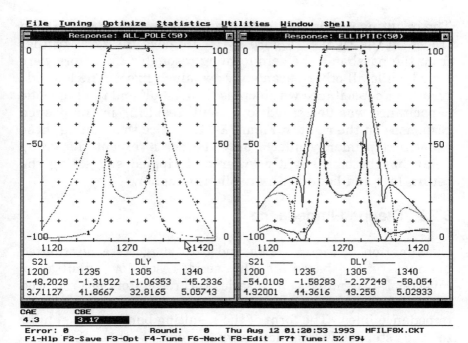

Figure 8-31 *On the left are amplitude and delay responses of a 7th order all-pole combline. On the right are responses with two zeros tuned closer (solid) and farther (dashed) from the passband.*

Closer spacing broadens the notches and wider spacings narrow the notches.

When the zeros are added they significantly perturb the return loss of the original filter. This is corrected by leaving the zero-forming resonator spacings and loading capacitors fixed and optimizing the all-pole parameters. The spacing between the end and the next inside resonator tends to decrease and the internal spacings increase. The loading capacitance on the end resonator adjacent to the low side zero-forming resonator increases and the loading capacitance on the opposite end decreases. The tap point moves higher up the resonators. The parameters after tuning and optimization are given in Table 8-14.

The resulting responses after optimization are given as the solid traces on the right in Figure 8-31. To move the zeros further from the passband, the zero forming resonator capacitors were tuned while all other parameters remained fixed. The low side resonator capacitance was increased to 4.3 pF and the high side capacitance was decreased to 3.17 pF resulting in the dashed responses on the right in Figure 8-31. Notice that tuning these capacitors primarily affects only the zeros, except the group delay has become slightly lower and rippled suggesting the return loss and bandwidth are slightly higher.

8.22   Stepped-Impedance

Examination of Figures 4-8a and 3-15 suggests a distributed bandpass filter based on the series-resonator admittance-inverter L-C bandpass. The series L-C resonators are replaced with high-impedance lines and the shunt capacitors with low impedance lines. The form of the resulting bandpass is similar to the stepped-impedance lowpass except that the high-impedance lines are just over 180 degrees long in the bandpass and as short as possible in the lowpass.

The stepped-impedance bandpass tends to be long because of the cascaded 180 degree resonators. Therefore this structure is most useful when a long and narrow aspect is desirable (such as coaxial filters), on substrates with a high dielectric constant to shorten the length, or at higher frequencies where the physical length is shorter. For example, a 7th order 880 MHz filter on PTFE board is over 48 inches long! The same filter at 10 GHz on alumina is 2 inches long.

The stepped-impedance bandpass has an additional lowpass passband as well as a reentrance mode at approximately two times the desired passband frequency. The ultimate rejection is therefore limited in the frequency regions between the passbands. This problem worsens with increasing bandwidth. For a typical 5th order Chebyshev with 40% bandwidth, the maximum rejection below the passband is 18 dB and the maximum rejection above the passband is 42 dB. However, at 5% bandwidth the same filter provides over a hundred decibels of ultimate rejection both above and below the passband.

At narrow bandwidth, realization issues become important. A 20% bandwidth filter can be realized with a high impedance line of 125 ohms and a low impedance line of 12.5 ohms, a ratio of ten to one. At 10% bandwidth, the low impedance line must be dropped to 11 ohms. By 5% bandwidth, the low-impedance line must be less than 6 ohms. These realizability issues and the ultimate rejection issue discussed above make this filter most suitable for moderate bandwidth applications.

Davis and Khan [14] give synthesis procedures which are more accurate than conversion of L-C filters designed by conventional admittance-inverter theory. This paper is directed at the coaxial process and consideration was given to compensating the coaxial steps, however certain concepts are useful for other processes as well. =M/FILTER= uses a unified technique to deal with step discontinuity absorption in all supported manufacturing processes.

Shown in Figure 8-32 is the layout and input parameter screen for a 10.7% bandwidth, 5.6 GHz, three section stepped-impedance bandpass filter on 25 mil thick PWB with a dielectric constant of 6.0. The cross hairs are 2.34 inches wide by 0.31 inches high. This length also includes 75 mil long 50 ohm leaders at each end of the filter. Also shown in Figure 8-32 is the =M/FILTER= View Electrical Variables window which gives the electrical parameters of the synthesized filter. L1 and L9 are the electrical length in degrees of high-impedance sections required by the first and last low-impedance impedance inverters. L2, L4, L6 and L8 are the electrical length in degrees for the low-impedance impedance inverter sections. L3, L5 and L7 are the lengths of the high-impedance resonator sections. All of these electrical lengths have been modified to compensate for effects of the width steps.

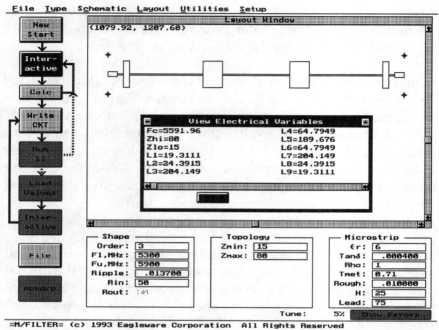

=M/FILTER= (c) 1993 Eagleware Corporation All Rights Reserved

Figure 8-32 *Layout and input parameter screen for a 5.6 GHz stepped-impedance bandpass. Also shown is a window with the electrical lengths of the lines.*

The simulator circuit file with a physical description of the filter is given in Table 8-15. After the file was written, it was manually modified to take advantage of structure symmetry and the narrow and wide line widths were changed to 12.5 and 200 mils from 13.1 and 206.2 mils respectively.

Given in Figure 8-33 on the left are the passband transmission and return loss amplitude responses before (solid) and after (dashed) optimization of the line lengths. On the right is the transmission amplitude for a broad sweep from dc to 14 GHz to show the lowpass passband, the first reentrance band and a portion of the second reentrance band.

Table 8-15 *Simulator circuit file with a physical description of the 5.6 GHz stepped-impedance bandpass.*

```
'   FILE: MFILF8W.CKT                    WI=36.7422
'   TYPE: Stepped - Bandpass - Microstrip  LI=75
'   FI: 5300 MHz   Fu: 5900 MHz          Whi=12.5
CIRCUIT                                  Wlo=200
SUB ER=6 TAND=0.0004 RHO=1               L1=?45.8971
& TMet=0.71 ROUGH=0.01 UNITS=0.0254      L2=?49.5358
MLI 1 2 W=WI H=H L=LI                    L3=?578.263
MST 2 3 O=SY NAR=WI W=Whi H=H            L4=?152.773
MLI 3 4 W=Whi H=H L=L1                   L5=?535.659
MST 4 5 O=SY NAR=Whi W=Wlo H=H           L6=L4
MLI 5 6 W=Wlo H=H L=L2                   L7=L3
MST 6 7 O=SY NAR=Wlo W=Whi H=H           L8=L2
MLI 7 8 W=Whi H=H L=L3                   L9=L1
MST 8 9 O=SY NAR=Whi W=Wlo H=H           WOUT=36.7422
MLI 9 10 W=Wlo H=H L=L4                  LOUT=75
MST 10 11 O=SY NAR=Wlo W=Whi H=H         WINDOW PASSBAND
MLI 11 12 W=Whi H=H L=L5                 FILTER(50,50)
MST 12 13 O=SY NAR=Whi W=Wlo H=H         GPH S21 -60 0
MLI 13 14 W=Wlo H=H L=L6                 GPH S11 -25 0
MST 14 15 O=SY NAR=Wlo W=Whi H=H         FREQ
MLI 15 16 W=Whi H=H L=L7                 SWP 4850 6350 151
MST 16 17 O=SY NAR=Whi W=Wlo H=H         OPT
MLI 17 18 W=Wlo H=H L=L8                 5300 5900 S11<-20
MST 18 19 O=SY NAR=Wlo W=Whi H=H         4850 5000 S21<-15
MLI 19 20 W=Whi H=H L=L9                 6200 6350 S21<-20
MST 20 21 O=SY NAR=Whi W=WOUT H=H        WINDOW REENTRANCE
MLI 21 22 W=WOUT H=H L=LOUT              FILTER(50,50)
DEF2P 1 22 FILTER                        GPH S21 -60 0
EQUATE                                   FREQ
H=25                                     SWP 0 14000 281
```

8.23 Stepped-Impedance BP Measured Responses

The 5.6 GHz stepped-impedance bandpass was etched on 25 mil thick, 0.5 ounce copper, Arlon GR6 with a specified relative dielectric constant of 6.0±0.25. SMA connectors were soldered directly to the PWB and the responses plotted in Figure 8-34 were measured. The center frequency is approximately 5553 MHz which is 0.84% lower than the design value. The bandwidth as defined by 15 dB return loss is approximately 655

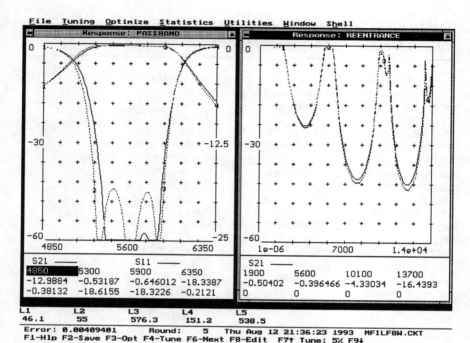

Figure 8-33 *Passband and stopband sweeps of the 5.6 GHz stepped-impedance bandpass filter.*

MHz which is 9.2% wider than the design value. These errors are most likely the combined result of etching and board thickness tolerance, a ±4.2% tolerance on the relative dielectric constant specification and simulator error modeling for the large step ratio (16:1).

8.24 Elliptic Direct Coupled

Elliptic function filters offer the promise of improved selectivity, particularly in transition region. However, two problems plague the application of elliptic transfer functions to distributed filters; a wide range of required line impedances and the difficulty of realizing certain resonator forms. These difficulties are partially managed using techniques by Rubenstein, et. al, [15] and Ness and Johnson [16].

S21 REF 0.0 dB 5.0 dB/   S11 REF 10.0 dB 5.0 dB/

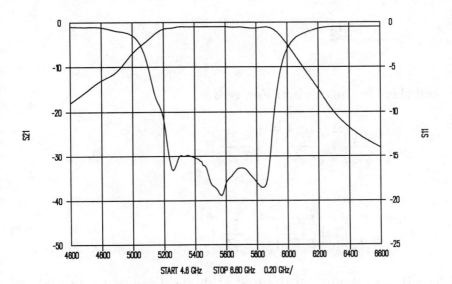

START 4.8 GHz   STOP 6.80 GHz   0.20 GHz/

Figure 8-34 *Measured responses of the 5.6 GHz stepped-impedance bandpass filter on 25 mil thick Arlon GR6.*

Shown in Figure 8-35 are the transformation steps from the elliptic lowpass prototype to the distributed elliptic bandpass. The elliptic lowpass prototype is first transformed into an elliptic bandpass (B) using the conventional transform. Next the Geffe transform is applied to the series branch transmission zero pairs to isolate the resonators (C). Admittance inverters are then applied to the parallel resonators in series with the transmission path to convert them to series resonators in shunt with the transmission path (D). Finally, the admittance inverters are realized as series lines and the resonators are realized as open stubs (E). The parallel-mode resonators are realized as half-wavelength open stubs and the series-mode resonators are realized as quarter-wavelength open stubs.

Design formulas are provided by Ness and Johnson and are not repeated here. However, the following expressions correct typographical errors in the original formula. First, in the original Figure 2

$$L'_1 = \frac{\phi L_1 L_2}{\phi L_2 - (1-\phi)L_1} \tag{25}$$

and then in the original Figure 3

$$J_{23} = \left(\frac{C_3 C'_2}{C'_3 L_2}\right)^{0.5} = \left(\frac{L'_3 C'_2}{L'_2 C'_3}\right)^{0.25} \left(\frac{2}{Z_{s2} Z_{s3}}\right)^{0.5} \tag{26}$$

$$J_{45} = \left(\frac{C_5 C'_4}{C'_5 L_4}\right)^{0.5} = \left(\frac{L'_5 C'_4}{L'_4 C'_5}\right)^{0.25} \left(\frac{2}{Z_{s4} Z_{s5}}\right)^{0.5} \tag{27}$$

The line to the left of the first stub and the line to the right of the last stub serve as impedance transformers. A higher internal impedance often improves the realizability of the stub and inverter line impedances.

The series of transformations involved in the creation of this filter, transmission line reentrance modes and the many discontinuities tend to obscure the true elliptic response. A_{min} is particularly susceptible to this problem while the passband characteristics are modified less. A successful design usually involves significant tuning and optimization to obtain a desired response. It is generally worthwhile to optimize an electrical description of the filter and use the optimized electrical values to prepare the physical description of the filter for final optimization of the physical parameters.

Elliptic filters in general, and this structure in particular, are more applicable for wider bandwidth applications. This structure has a number of useful design freedoms which assist with managing realizability. However, below about 15%

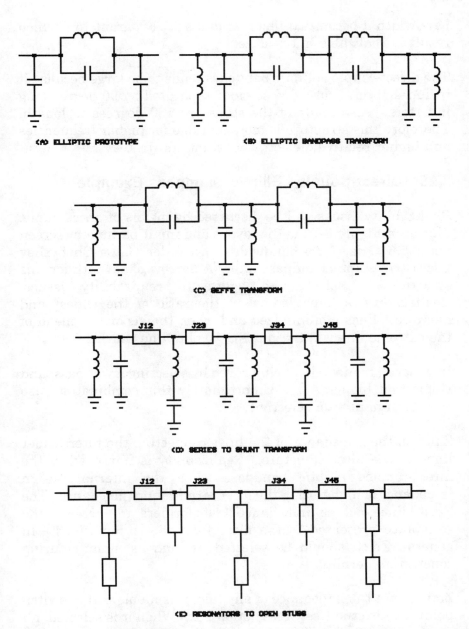

(A) ELLIPTIC PROTOTYPE

(B) ELLIPTIC BANDPASS TRANSFORM

(C) SEFFE TRANSFORM

(D) SERIES TO SHUNT TRANSFORM

(E) RESONATORS TO OPEN STUBS

Figure 8-35 *Transformations from the elliptic lowpass prototype (A) to the distributed elliptic bandpass (E). All line elements in (E) are 90 degrees long.*

bandwidth it becomes difficult to find a set of parameters which result in realizable line widths.

The direct-coupled elliptic bandpass tends to be large because it includes several cascaded sections separated by 90 degree long inverters. Also many of the stubs are 180 degrees in length. Therefore this structure is more suitable for higher frequencies and higher relative dielectric constant substrates.

8.25   Direct-Coupled Elliptic Bandpass Example

Realizability issues will be discussed by means of an example. Shown in Figure 8-36 is the layout and input parameter screen for a 5.6 GHz, 7.2% bandwidth, 3rd order Cauer-Chebyshev microstrip elliptic bandpass filter. A narrow bandwidth for this structure is selected to aggravate realizability issues. Realizability is improved when the ratio of the widest and narrowest lines is minimized and when the geometric mean of the extreme line widths equals a 50 ohm line width.

In general, realizability is improved by selecting lower passband ripple and higher A_{min}. Unfortunately this combination also results in minimum selectivity.

Zmch is the impedance of the lines connecting the filter leader lines to the filter structure. Realizability is improved in the direct-coupled elliptic bandpass when the internal design impedance is higher than the filter termination impedance. The Zmch lines act as impedance transformers to increase the resistance presented to the filter from the terminations. In general, Zmch should be selected as high as manufacturing constraints permit.

Zmin is not the impedance of any line; it is an algorithm switch point. Notice the Use Cross option box. When it is selected, all stubs with impedances which drop below Zmin are converted to double stubs which are inserted in the filter with cross models. The width of a double stub is narrower than an equivalent single

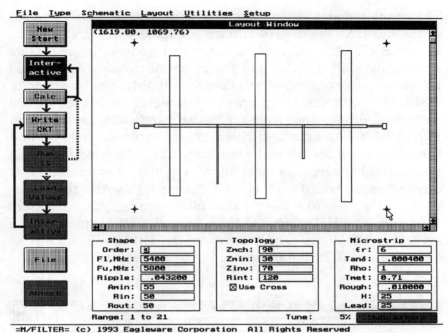

Figure 8-36 *Layout and input parameter screen for the 5.6 GHz microstrip elliptic bandpass filter.*

stub. Double stubs shorten the length of the filter slightly but they increase the width significantly. Narrower stub line widths improves simulation accuracy.

Zinv is the impedance of the 90 degree long admittance-inverter sections along the length of the filter. Zinv should be a moderate impedance, but higher than the termination resistance. Typical values are 60 to 90 ohms for 50 ohm terminations.

Rint is also not a line impedance but an internal filter impedance. After other parameters are selected, Rint is adjusted for best realizability. Typical values are 70 to 200 ohms for 50 ohm terminations.

The simulated amplitude transmission and return loss responses of the 5.6 GHz microstrip elliptic bandpass filter are given in Figure 8-37. The solid traces are with physical dimensions as written by =M/FILTER= and before optimization. The dashed traces are after optimization. Some manual tuning was used to assist the optimization process. Simulator responses of the optimized filter swept over a broader bandwidth are given in Figure 8-38. As expected, the numerous transformations and the behavior of distributed resonators have obscured $A_{min.}$ The transmission zeros of an ideal equally terminated L-C conventional transform elliptic bandpass with the filter parameters used to generate this filter occur at approximately 4200 and 7500 MHz. Zeros at these approximate frequencies are evident in the distributed filter response, but significant modes occur further from the passband in the distributed filter.

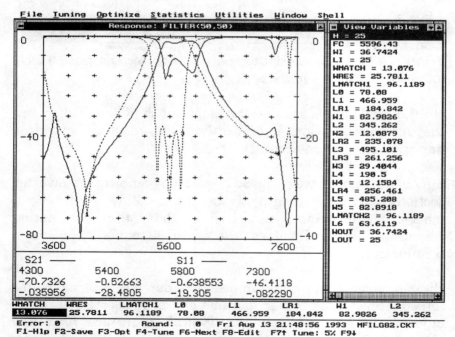

Figure 37 Amplitude transmission and return loss responses of the elliptic bandpass before (solid) and after optimization (dashed). Optimized dimensions are given on the left.

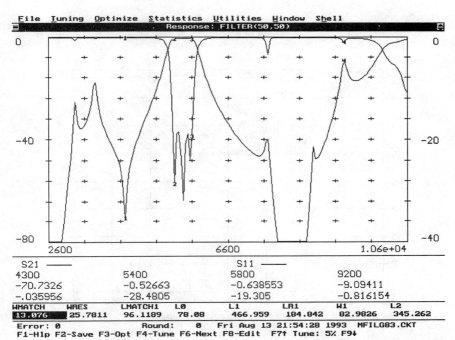

Figure 8-38 *Broadband sweep of the 5.6 GHz elliptic bandpass filter.*

8.26 Elliptic Bandpass Measured Data

The 5.6 GHz microstrip direct-coupled elliptic bandpass was etched on 25 mil thick, 0.5 ounce copper, Arlon GR6 with a specified relative dielectric constant of 6.0±0.25. SMA connectors were soldered directly to the PWB and the responses plotted in Figure 8-39 were measured. In general, the passband characteristics are as expected. The return loss is worse than expected. If required, trimming of the stub lengths would probably recover the response. The 180 degree stubs are associated with the passband parallel resonators and they are the more likely candidates for trimming. The stopband performance is also not as good as the simulation predicts, particularly in the frequency region from 7.5 to 9.5 GHz. This is possibly the result of surface waves which could be suppressed

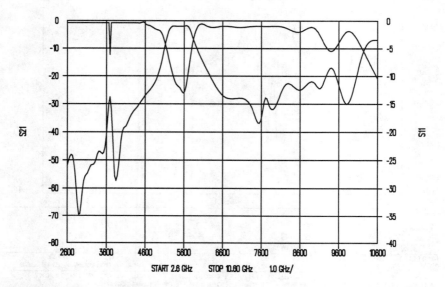

S21 REF 0.0 dB 10.0 dB/　　S11 REF 0.0 dB 5.0 dB/

START 2.8 GHz　　STOP 10.80 GHz　　1.0 GHz/

Figure 8-39 *Measured transmission amplitude and return loss responses of the microstrip 5.6 GHz elliptic bandpass on 25 mil thick Arlon GR6 with a relative dielectric constant of 6.0±0.25.*

by placing the PWB in a housing below cutoff. However, the physical width of this structure with just over 1/8 inch PWB edge clearance is 1.3 inches which has a housing cutoff frequency of 4.5 GHz. The single stub form is narrow enough to be placed in a housing with a cutoff which would improve rejection in the lower portion of the 7.5 to 9.5 GHz band. This points to the advantage of physically small filter structures. For example, if this filter were constructed on alumina with a relative dielectric constant of 9.6, the single stub form would be sufficiently narrow to be placed in a housing below cutoff though the entire sweep band above.

An HPGL file layout with final dimensions generated by =M/FILTER= and imported to the publishing system for this book is given in Figure 8-40. The scale is approximately 2:1 and there is a large 5 mil etch factor.

8.27 Evanescent Mode Waveguide Filters

Waveguide below cutoff does not propagate. Signals die exponentially with length, thus the term evanescent mode. At frequencies well below cutoff, the attenuation is frequency independent. Mechanically variable attenuators were the primary application of below cutoff waveguide until it was discovered that evanescent mode guide behaves like a pi or tee of inductors. Examination of Figure 4-5b suggests that a pi of inductors could serve as a bandpass filter provided capacitive elements are suitably placed to load the guide. These concepts are introduced in Sections 3.35 and 3.36 where expressions are given for the inductive reactance of the pi and tee models of

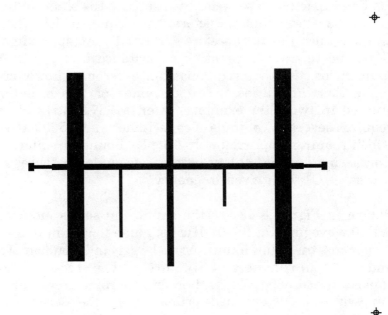

Figure 8-40 *Layout of the 5.6 elliptic bandpass. The scale is approximately 2:1.*

evanescent mode waveguide. In this chapter, we will use these concepts to design evanescent mode bandpass filters.

Equations (3-121,122) were used to find the reactance of the shunt inductors in the evanescent mode pi equivalent model of X-band WR90 rectangular copper waveguide (a=900, b=400 mils) for operating frequencies from 250 to 6557 MHz. As the length of guide is increased the series inductive reactance becomes large (narrow bandwidth filters) and the shunt inductive reactance is independent of the length. The shunt reactance was converted to effective inductance and plotted in Figure 8-41. Well below cutoff the inductance asymptotically approaches 4.115 nH. As the frequency approaches cutoff at 6557.22 MHz, the shunt inductance becomes infinite. Also plotted in Figure 8-41 is the capacitance which resonates the shunt inductance at a given frequency on the independent axis. When the evanescent mode waveguide sections are placed end to end with loading capacitance between each section, the shunt inductance is two parallel shunt inductors, each equal to L1. The total shunt inductance is therefore 1/2 of L1. At approximately 3 GHz, the required resonating capacitance in 1.0 pF. At low frequencies, the required capacitance becomes large, creating realization difficulties. Two extremes of this situation are studied in two filter examples later, a 987.5 MHz filter with requires several picofarads of capacitance and a 5600 MHz filter which requires approximately 0.1 pF. Round or square tubing (increasing b until b=a), or larger waveguide, increases L1 and decreases C1 for a given frequency.

Shown in Figure 8-42 are the shunt and series inductance of WR90 waveguide at 3000 MHz vs. guide length in inches. For long guide, the shunt inductance is largely independent of length and is approximately 4.574 nH. The required loading capacitance at 3000 MHz is therefore approximately 1 pF. Filter bandwidth is determined primarily by the series coupling inductance L3. Decreasing bandwidth requires larger inductance and therefore longer guide sections.

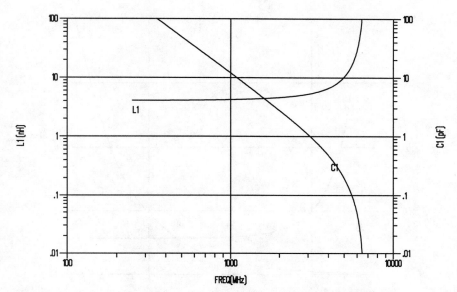

Figure 8-41 *Shunt inductance, L1, of WR90 waveguide vs. frequency below cutoff (evanescent mode) and the required capacitance to resonate L1 at the frequency of the independent axis.*

The required length of each section is given approximately by [17]

$$l = \frac{1}{\gamma} \cosh^{-1} \frac{\Delta}{bw} \tag{28}$$

where bw is the fractional bandwidth, γ is defined in Section 3.35, and Δ is given by

$$\Delta = \frac{2}{1 + \dfrac{1}{1 - (\lambda_c/\lambda_o)^2}} \tag{29}$$

For operation at frequencies well below the cutoff, $\Delta=1$. At 3000 MHz in WR90 waveguide, $\Delta=0.883$.

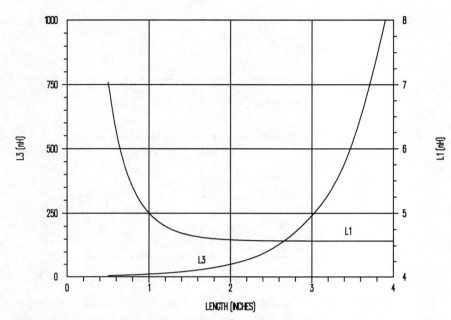

Figure 8-42　*Evanescent mode waveguide equivalent shunt inductance and series inductance at 3000 MHz vs. guide length in inches for WR90.*

8.28　Evanescent Mode Loading Capacitance

A short section of guide loaded with dielectric material forms a capacitive element. This is often used for wideband evanescent mode filters. For narrowband filters, a post perpendicular to the broad dimension and which approaches the opposite wall forms a parallel plate capacitor, C_p. A typical post diameter is one tenth the broad dimension of the waveguide. An estimate of C_p is given by equation (3-29). Additional fringing capacitance, C_f, from the post to the side walls is approximately

$$C_f = 0.68d \tag{30}$$

where C_f is in picofarads, d is the diameter of the post in inches and $d/a < 0.5$. The fringing capacitance of a post with a diameter of one tenth the broad dimension in WR90 guide is therefore only 0.061 pF. The total capacitance, equal to $C_p + C_f$, is therefore

$$C_t = \frac{0.2248 \, \epsilon_r \, A}{s} + 0.68d \tag{31}$$

where all dimensions are in inches. A is the area of the post end and s is the spacing between the post and the guide wall.

The inductance of the post cancels a portion of the capacitive reactance. This is compensated for by tuning the post length. Caution should be exercised when using conventional waveguide references for capacitive elements because discontinuities in waveguide below cutoff behave differently than discontinuities in propagating guide.

It is difficult to overemphasize the importance of recognizing that evanescent mode filters behave precisely like lumped element filters. For example, the capacitive elements may literally be lumped elements such as commercial piston trimmers. It is also feasible to increase the capacitance by having the post penetrate a well hole to form a coaxial capacitor. These techniques are considered by Snyder [18]. The unloaded Q of an evanescent mode resonator may be severely degraded by the unloaded Q of the capacitive loading element.

8.29 Coupling to Evanescent Mode Waveguide

Evanescent mode filter resonators and internal resonator coupling are defined above. It remains to discuss methods of external coupling. Depicted in Figure 8-43 are four possible external coupling methods. For the top filter, a conducting post extends from the center pin of the TEM mode connector and contacts the opposite conducting wall of the guide. A schematic representation of this configuration is given on the top in Figure 8-44. The coupling post is in parallel with the shunt inductors

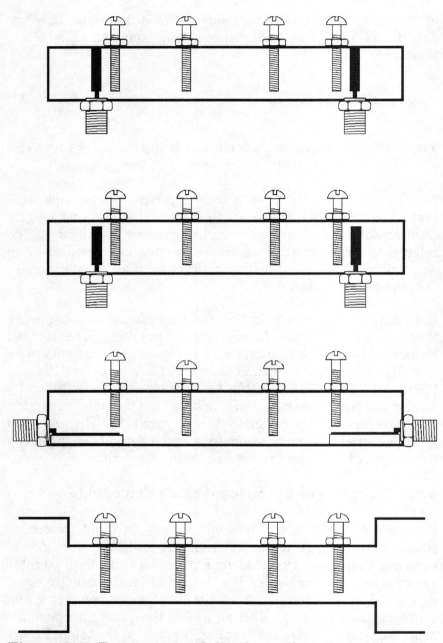

Figure 8-43 *Evanescent mode bandpass filters depicting various external coupling methods. See text for details.*

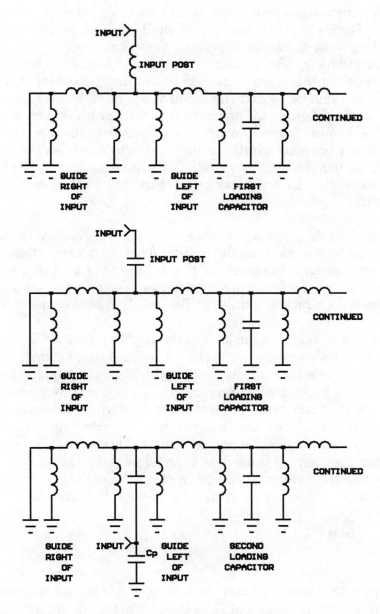

Figure 8-44 *Schematic representation of the top three evanescent mode filter external coupling methods depicted in Figure 8-43.*

of the evanescent mode waveguide on both sides of the post. From Figure 8-42 we see that if the length of guide from the coupling post to the first loading capacitor is short, then the shunt inductor, L3, of that section is large and unimportant. The input is therefore connected to the first resonator through the post inductance and the small value of series inductor L1. The section of guide to the left of the input post is shorted at its left end and inspection of the schematic reveals that it effectively presents additional inductance to ground at the input. Later we use these models to simulate the response of the entire filter structure by computer and to further investigate external coupling.

The second diagram in Figure 8-43 depicts capacitive coupling into the evanescent mode filter. This structure offers the opportunity for adjustment of the coupling via a tuning screw. The electrical configuration is given as the second schematic in Figure 8-44 which is similar to the coupling post configuration.

The third evanescent mode external coupling method in Figure 8-43 was discussed by Bharj [19]. The TEM input connects to a microstrip line on a substrate placed within the guide which is modeled as capacitance to ground, C_p. This line is coupled to the evanescent mode waveguide via a capacitive post similar to a post used to form the resonators. A schematic for this configuration is given at the bottom of Figure 8-44. When the microstrip length is much less than 90 degrees, the capacitance to ground, C_p, is estimated using a parallel plate capacitor model (equation 3-29) or as an open transmission line stub using

$$C_p = \frac{\tan\theta}{\omega Z_o} \tag{32}$$

where ω is the filter center frequency, θ is the electrical length of the microstrip line and Z_o is the characteristic impedance of the line.

The final configuration depicted in Figure 8-43 couples to the evanescent mode waveguide by a sudden step to propagating waveguide. This is discussed by Craven and Mok [17] and Snyder [18]. The capacitive susceptance of the step junction is compensated by a short length of evanescent mode waveguide between the propagating guide and the first loading capacitor. Design details are given in the cited references. Other propagating guide coupling techniques have been used such as dielectric or ridge loading of the small evanescent mode guide to reduce the cutoff frequency below the operating frequency. Gradual tapers do not work because the impedance varies over a range of both real and imaginary values through the taper.

8.30   Reentrance in Evanescent Mode Filters

Spurious reentrance modes occur in evanescent mode filters at frequencies where the guide wavelength between the capacitive discontinuities is one-half wavelength. These frequencies must be at least higher than the dominant mode cutoff frequency. The bandwidth of the desired passband controls the spacing between the loading capacitors with narrow bandwidths having the largest spacings and therefore the lowest reentrant frequencies. Operation well below cutoff can insure stopband performance of arbitrary width at the expense of resonator unloaded Q and large loading capacitance requirements. In general, practical evanescent mode filters have stopband performance at least equal to, or superior to, other distributed filter structures. Specific reentrance performance is considered in the following evanescent mode filter examples.

8.31   996 MHz Evanescent Mode Filter Example

The first evanescent mode bandpass filter example is a three section 996 MHz filter with the structure depicted at the top of Figure 8-43. The desired bandwidth is 32 MHz (a fractional bandwidth of 0.0321). One inch square aluminum extruded tubing with inside dimensions of 875×875 mils is used for the waveguide. The guide was left open on the ends as opposed to shorted as in Figure 8-43. The cutoff frequency of the 875 mil tubing is 6745 MHz. From equation (3-118), γ is 3.551 inches and from equation (28), Δ is 0.989. Therefore, from equation (29), the estimated spacing between loading capacitors is 1.16 inches. From equations (117) and (121), the inductive reactance is 58.15 ohms and the effective inductance at 996 MHz is 9.291 nH. The approximate loading capacitance to resonate two 9.291 nH inductors in parallel is 5.5 pF. The final loading capacitance will be somewhat larger than this because the series inductances reduce the effective inductance at the resonant node.

A circuit file for this filter is given in Table 8-16. The inductance variables LPE and LSE are the pi model parallel (shunt) and series inductance for the evanescent mode section of waveguide left of the input post. LP1, LP2 and LP3 are the pi model parallel inductances of the first, second and middle filter sections. The circuit file takes advantage of symmetry by using the first two sections for the output sections. The inductances are defined within the circuit file using the evanescent mode waveguide equations given in Section 3.25. This provides for optimization of physical dimensions of the evanescent mode filter. A and B are the waveguide inside dimensions in mils. L1, L2 and L3 are the length of the guide sections. From these, the cutoff wavelength, γ (GAMMA) and characteristic reactance (X) and finally the pi model inductances are found. An advantage of this approach is that the frequency dependent behavior of the model is predicted and the method is accurate for wide bandwidth filters. The responses for this filter after optimization of the section lengths and the loading capacitances is given in Figure 8-45.

The loading capacitances C1 and C2 are too large to realize using a capacitive post. Commercial glass piston trimmers of sufficient length to extend wall to wall were placed in the guide. The diameter of these trimmers is approximately 0.25 inches which is 29% of the guide width, much large than the recommended 10% post diameter. This leads to ambiguity in the correct physical spacing. The lengths L1 and L2 which resulted

Table 8-16 *Simulator circuit file for the 996 MHz evanescent mode filter in 875 mil square guide.*

```
'EVANESCENT MODE FILTER            LAMBAIR=11803000/FREQ
'996 MHz 3 SECTION PLUS END        LAMBCUT=2*A
LAUNCHING                          GF=SQR((LAMBAIR/LAMBCUT)^2-1)
'875 X 875 MIL TUBING              GAMMA=2*PI*GF/LAMBAIR
CIRCUIT                            X=120*PI*B/A/GF
IND 5 0 L=LPE Q=QR                 LPE=X/TANH(GAMMA*LE/2*57.3)
IND 2 5 L=LSE Q=QR                 LPE=LPE*1E3/(2*PI*FREQ)
IND 2 0 L=LPE Q=QR                 LSE=X*SINH(GAMMA*LE*57.3)
IND 1 2 L=LC  Q=QR                 LSE=LSE*1E3/(2*PI*FREQ)
IND 2 0 L=LP1 Q=QR                 LP1=X/TANH(GAMMA*L1/2*57.3)
IND 2 3 L=LS1 Q=QR                 LP1=LP1*1E3/(2*PI*FREQ)
IND 3 0 L=LP1 Q=QR                 LS1=X*SINH(GAMMA*L1*57.3)
CAP 3 0 C=C1 Q=QC                  LS1=LS1*1E3/(2*PI*FREQ)
IND 3 0 L=LP2 Q=QR                 LP2=X/TANH(GAMMA*L2/2*57.3)
IND 3 4 L=LS2 Q=QR                 LP2=LP2*1E3/(2*PI*FREQ)
IND 4 0 L=LP2 Q=QR                 LS2=X*SINH(GAMMA*L2*57.3)
CAP 4 0 C=C2 Q=QC                  LS2=LS2*1E3/(2*PI*FREQ)
DEF2P 1 4 HALF                     LP3=X/TANH(GAMMA*L3/2*57.3)
HALF 1 2 0                         LP3=LP3*1E3/(2*PI*FREQ)
IND 2 0 L=LP3 Q=QR                 LS3=X*SINH(GAMMA*L3*57.3)
IND 2 3 L=LS3 Q=QR                 LS3=LS3*1E3/(2*PI*FREQ)
IND 3 0 L=LP3 Q=QR                 WINDOW
HALF 4 3 0                         EVANESC(50)
DEF2P 1 4 EVANESC                  GPH S21 -80 0
EQUATE                             GPH S11 -30 0
A=875                              SMH S11
B=875                              MARKER
QR=4000                            932 980 1012 1060
QC=165                             FREQ
LC=4                               SWP 916 1076 81
C1=?5.836026 '5.5                  OPT
C2=?5.6839 '5.5                    980 1012  S11<-18
L1=?320 '450                       916 932   S21<-46 W21=1000
L2=?1005 '1045                     1060 1076 S21<-46 W21=1000
L3=?1085 '1125
LE=500
```

in an optimum simulated response were 1005 and 1085 mils respectively. Recall the approximate spacing predicted by equation (29) was 1160 mils. The input coupling post was also 0.25 inches in diameter. The spacing from the coupling post to

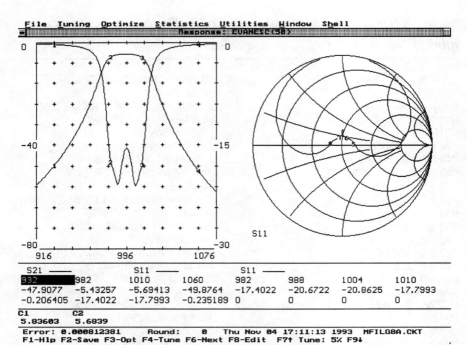

Figure 8-45 *Responses of the 996 MHz evanescent mode waveguide bandpass filter in 875 mil square aluminum tubing.*

the first loading capacitor is L1. The following measured responses are for a filter with L1=450 mils, L2=1045 and L3=1125. The simulated responses require smaller spacings as indicated in the circuit file. The differences are likely due to the large diameter of the posts and trimmers which results in series susceptances associated with these obstacles.

Measured responses of the 996 MHz evanescent mode filter are given in Figure 8-46. The insertion loss at band center is approximately 4.4 dB. The theoretical unloaded Q of 875 mil square aluminum tubing at 996 MHz is 4270. QR in the circuit file was set at 4000. This required that the unloaded Q for the piston trimmer loading capacitors be set at 165 if the simulated loss was to match the measured loss. This rather disappointing unloaded Q is largely a result of the fact that the operating frequency is 6.77 times lower than the cutoff frequency. The required loading capacitance was therefore high and the capacitive reactance is approximately 27.5 ohms. The effective series RF loss resistance which results in an unloaded Q of 165 for the trimmers is only .17 ohms.

8.32   5.6 GHz Evanescent Mode Filter Example

The second evanescent mode filter example is a 3-section 5600 MHz bandpass filter in 750×375 mil copper waveguide. An unloaded Q of the evanescent mode guide as predicted by equation (3-124) is approximately 6400. The required loading capacitance is far less than the previous evanescent mode example because the operating frequency is only a factor of 1.4 lower than the cutoff frequency (7869 MHz) and the operating frequency is a factor of 5.6 higher than the previous filter.

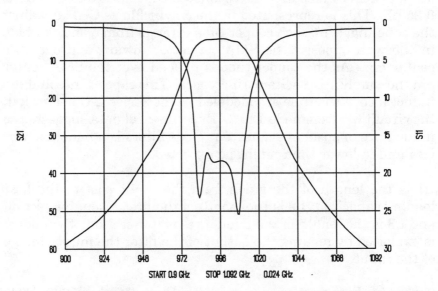

Figure 8-46 *Measured responses of the 996 MHz evanescent mode filter in 875 mil square aluminum tubing with with piston trimmer loading capacitance.*

A circuit file is given in Table 8-17. As with the previous example, dimension variables are used in the circuit file which are converted into inductance model values using the equations given in Section 3.35. LE is the length of the guide from the input post to the ends of the guide. The end of the guide is terminated in RE which is set at a low value to simulate a short. However, changing RE to a high value has little effect on the simulated response indicating the length of guide, LE, is sufficiently long to isolate the filter ends from the design. Tests with the final filter indicated that whether the ends are open or closed has little effect on the passband response. However, open ends increased the passband loss slightly and decreased stopband attenuation, indicating a small amount of signal leaks from an open end and couples to the opposite end.

The coupling structure used was similar to the bottom schematic of Figure 8-44 except the PWB was replaced with a 100 mil extension into the guide of the center pin of the input and output SMA connector. The estimated capacitance to ground is 0.25 pF. This is represented in the circuit file as CG. To adjust the coupling, a 1/8 inch copper tube extends from a snug-fit hole in the wall opposite the SMA connector toward and over the center pin. As the copper tube is pushed over the SMA center pin the coupling capacitance increases. This capacitance and the inductance of the tube are modeled as the series L-C network in the circuit file from node 1 to 2. Experimental data suggests the inductance of posts in evanescent mode filters is a nanohenry or less and is lower than might be expected.

L1 is the length of the guide from the input post to the first loading capacitor. L2 is the length of the first resonator section and L3 is the length of the middle resonator section. Symmetry is exploited by making the output of the filter the mirror image of the input.

Given in Figure 8-47 are are responses of the 5600 MHz evanescent mode filter as computed by =SuperStar= Professional. Also shown in a window next to the responses are

Table 8-17 *Circuit file for a 3-section 5600 MHz evanescent mode bandpass filter with inductive coupling posts in 750×375 mil copper waveguide.*

```
'EVANESCENT MODE FILTER          LAMBAIR=11803000/FREQ
'5550-5650 MHz                   LAMBCUT=2*A
'3 SECTION PLUS ENDS             GF=SQR((LAMBAIR/LAMBCUT)^2-1)
'750 X 375 GUIDE                 GAMMA=2*PI*GF/LAMBAIR
CIRCUIT                          X=120*PI*B/A/GF
RES 8 0 R=RE                     LSE=X*TANH(GAMMA*LE/2*57.3)
IND 8 7 L=LPE                    LSE=LSE*1E3/(2*PI*FREQ)
IND 7 0 L=LPE                    LPE=X/SINH(GAMMA*LE*57.3)
IND 7 2 L=LSE                    LPE=LPE*1E3/(2*PI*FREQ)
CAP 1 0 C=CG                     LS1=X*TANH(GAMMA*L1/2*57.3)
SLC 1 2 L=LC C=CC                LS1=LS1*1E3/(2*PI*FREQ)
IND 2 3 L=LS1 Q=6460             LP1=X/SINH(GAMMA*L1*57.3)
IND 3 0 L=LP1 Q=6460             LP1=LP1*1E3/(2*PI*FREQ)
IND 3 4 L=LS1 Q=6460            LS2=X*TANH(GAMMA*L2/2*57.3)
CAP 4 0 C=C1 Q=Qc               LS2=LS2*1E3/(2*PI*FREQ)
IND 4 5 L=LS2 Q=6460            LP2=X/SINH(GAMMA*L2*57.3)
IND 5 0 L=LP2 Q=6460            LP2=LP2*1E3/(2*PI*FREQ)
IND 5 6 L=LS2 Q=6460            LS3=X*TANH(GAMMA*L3/2*57.3)
CAP 6 0 C=C2 Q=Qc               LS3=LS3*1E3/(2*PI*FREQ)
DEF2P 1 6 HALF                   LP3=X/SINH(GAMMA*L3*57.3)
HALF 1 2 0                       LP3=LP3*1E3/(2*PI*FREQ)
IND 2 3 L=LS3 Q=6460            WINDOW
IND 3 0 L=LP3 Q=6460            EVANESC(50)
IND 3 4 L=LS3 Q=6460            GPH S21 -50 0
HALF 5 4 0                       GPH S11 -25 0
DEF2P 1 5 EVANESC               MARKER
EQUATE                          5550 5600 5650 5800
Qc=900                          FREQ
A=750                           SWP 5350 5850 201
B=375                           OPT
RE=1E-06                        5550 5650 S11<-15
LC=0.5                          5350 5400 S21<-56 W21=1E5
CC=?0.1044457                   5800 5850 S21<-50 W21=1E5
CG=0.25
C1=?0.2714525
C2=?0.302066
L1=240
L2=1315
L3=1380
LE=685
```

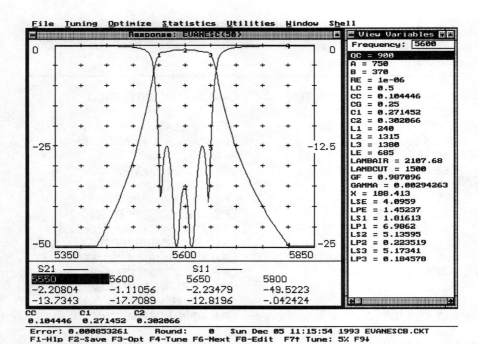

Figure 8-47 *Responses and variable values for the 5600 MHz 3-section evanescent mode bandpass filter.*

the variable values defined and computed in the EQUATE block of the circuit file. The measured responses are given in Figure 8-48. The inductors representing the evanescent mode guide are set to an unloaded Q of 6400. The capacitor unloaded Q was adjusted until the predicted loss matched the measured loss. With brass 6-32 screws for the capacitors, the capacitor unloaded Q which resulted in the measured loss of approximately 1.1 dB was 900. A center frequency of 5600 MHz required a screw penetration within approximately 100 mils of the waveguide floor. Next, the threads of that portion of the screws which penetrated the waveguide were turned smooth. The new surface was plated with a thin layer of silver. The loss was reduced to just under 1 dB which is simulated with an unloaded Q of 1100. The measured bandwidth after removing the threads from the tuning screws was narrower by a few percent, indicating the the diameter of the capacitive posts modifies the effective length of

Bandpass Structures 379

S21 REF 0.0 dB 10.0 dB/   S11 REF 0.0 dB 5.0 dB/

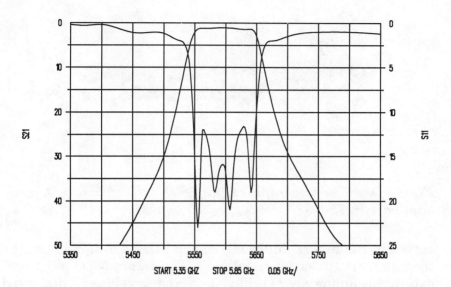
START 5.35 GHZ   STOP 5.85 GHz   0.05 GHz/

Figure 8-48 *Measured responses of the 5600 MHz evanescent mode bandpass in 750×375 mil waveguide.*

the guide sections.

A photograph of the 5600 MHz evanescent mode bandpass is given in Figure 8-49. The copper tubes used for adjustment of the series coupling capacitance are seen extending outside the wall opposite the SMA input and output connectors. The tuning screws are also on the wall opposite the connectors. Springs are placed between the threaded bushings and the heads of the tuning screws to keep the screws snug during tuning. A more permanent arrangement after tuning is to use lock nuts on the tuning screws.

8.33 Filters With Arbitrary Resonator Structure

As was introduced in Section 4.23, bandpass filters may be defined by only three entities; the resonator structure, coupling

Figure 8-49 *Photograph of the 5600 MHz evanescent mode bandpass in 750×375 mil copper waveguide.*

between resonators (internal coupling) and coupling to the terminations (external coupling). The internal coupling and external coupling are specified as K and Q values as discussed in Section 4.23. In this section, we will use an experimental procedure to illustrate the design of nearly arbitrary bandpass filter structures using the K and Q values.

The resonator form which is selected for a first example lends itself to quick construction. A typical application might be a single filter for lab use which needs to be built in a few hours. A sketch of the resonator form is given at the bottom of Figure 8-50.   The length of wire from the center of the wire at the grounded end to the loading capacitor connection point is approximately two inches. The center of the wire is 0.255 inches above the ground plane.   The capacitor is adjusted to set resonance at 680 MHz.

A sketch of a 5-section filter using this resonator is given at the top of Figure 8-50. The structure is a combline and the resonators couple to each other because of their proximity. Conventional combline design theory can not be used because coupled wire over ground data is unpublished, the bends at the

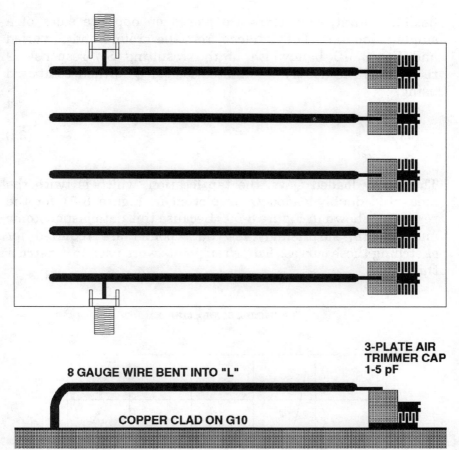

3-PLATE AIR TRIMMER CAP 1-5 pF

8 GAUGE WIRE BENT INTO "L"

COPPER CLAD ON G10

Figure 8-50 *Resonator formed by a length of 8 gauge wire over a ground plane loaded and tuned with a 3-plate silver plated air trimmer capacitor.*

ground end would modify this data even if it existed and the coupling between loading capacitors is unknown. External coupling is via a tap of the end resonators to an SMA connector.

To complete the design, it is necessary to determine the required spacings between resonators and the external tap location. Considering all of the above factors, an analytical solution is intractable, so we will use an empirical method which is very flexible. First, connectors are placed on opposite sides of a single resonator. The distance from the ground end is varied and the 3 dB bandwidth of the resulting S_{21} response is recorded. The doubly terminated loaded Q is then computed using

$$Q_d = \frac{f_o}{BW} \tag{33}$$

The singly loaded Q vs. the tap location, which is twice the measured doubly-loaded Q, is plotted in Figure 8-51 for the resonator shown in Figure 8-50. Because this data is monotonic and smooth, surprisingly few data points are required for sketching these curves. Three tap points were used to construct this data.

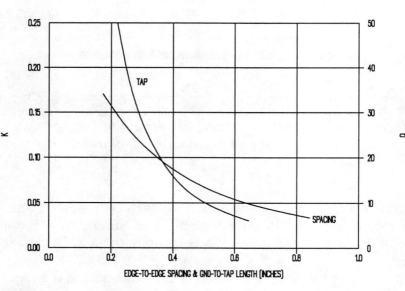

MEASURED EXTERNAL AND INTERNAL COUPLING DATA

Figure 8-51 *Internal coupling vs. edge-to-edge wire spacing and resonator loaded Q vs. length of tap from the ground end for the resonator form shown in Figure 8-50.*

Also plotted in Figure 8-51 is the internal coupling coefficient K vs. the edge-to-edge wire spacing between two resonators. This data is obtained by varying the spacing between two resonators and measuring S_{11} while very lightly coupling to the two resonators. In this case, the SMA connectors with an open pin were placed approximately 0.5 inches outside a respective resonator. A sweep of S_{11} reveals high reflection except at two frequency points where notches occur. The coupling coefficient K is then found from

$$K = \frac{2(f_2 - f_1)}{f_2 + f_1} \tag{34}$$

where f_1 and f_2 are the lower and upper notch frequencies in the S_{11} responses. Again, because K vs. spacing is monotonic, only a few data points are required. In practice, the range of spacings over which data is to be taken is determined by completing the following steps first.

We desire a 5-section 0.177 dB ripple Chebyshev 650 to 710 MHz bandpass. The lowpass prototype g values are $g_o=1$, $g_1=g_5=1.3014$, $g_2=g_4=1.3457$ and $g_3=2.1281$. Using equations (4-38) through (4-40), the normalized internal coupling values are $k_{12}=k_{45}=0.7557$ and $k_{23}=k_{34}=0.05909$ and the normalized loaded Qs are $q_1=q_n=1.3014$. These values are denormalized to the desired filter bandwidth using equations (4-41) through (4-43). Therefore

$$Q_1 = Q_n = \frac{1.3014 \times 680}{60} = 14.75 \tag{35}$$

$$K_{12} = K_{45} = \frac{0.7557 \times 60}{680} = 0.0667 \tag{36}$$

$$K_{23} = K_{34} = \frac{0.5909 \times 60}{680} = 0.0521 \tag{37}$$

From Figure 8-51 we find that $K=0.0667$ requires a wire spacing of 0.51 inches which is then used for the spacing between the first and second and the fourth and fifth resonators. $K=0.0521$ requires a spacing of 0.62 inches which is the spacing between the second and third and the third and fourth resonators. Also from Figure 8-51, we find that a loaded Q of 14.75 requires a tap location of approximately 0.41 inches from the ground. The filter was constructed with these dimensions and the responses shown in Figure 8-52 were measured after the trimmer capacitors were adjusted for best return loss.

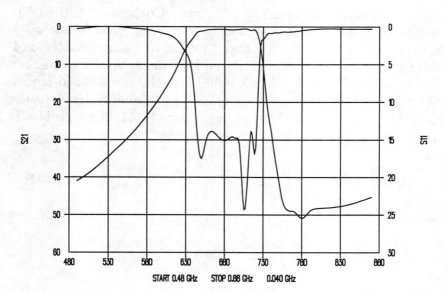

Figure 8-52 *Responses of the wire over ground combline bandpass designed using a flexible empirical method.*

Notice that the bandwidth is significantly wider than desired. Also, there is a notch above the passband which results in greater selectivity on the high side than would be expected from a 5-section filter. The empirical design method precisely models a single resonator and resonator pairs, and therefore the suspect cause for both response discrepancies is coupling between nonadjacent resonators. It is the author's experience that combline structures with a conductor diameter small in relation to the ground spacing are more susceptible to nonadjacent resonator coupling.

One possible method to correct the bandwidth is to redesign for a narrower bandwidth using the same Figure 8-51. The resulting bandwidth would then be closer to the desired bandwidth. However, a second approach was selected. A shield is placed one inch above the ground plane which naturally decreases the adjacent resonator coupling. The data in Figure 8-51 is no longer valid, however, since the expected effect of the shield is to decrease the bandwidth, the experiment is justified.

The results after placing the shield over the resonators and retuning the trimmer capacitors is given in Figure 8-53. The bandwidth is in fact narrower and closer to the desired value. The return loss is degraded probably due either to a need to readjust the tap point or a need to regenerate Figure 8-51 and recalculate the spacings between resonators. Also, the insertion loss of the shielded filter is about a tenth of a decibel less than the unshielded filter even though the shielded filter has a narrower bandwidth. This suggests that the unshielded filter radiates slightly.

8.34 Hidden Dielectric Resonator Example

Conventional PWB materials, such as G-10 and FR4, are inexpensive and readily available. However, the dielectric stability and loss tangent properties are poor. Figure 3-32 illustrates that these materials cannot provide unloaded Qs much higher than 100. In this example, we will design a

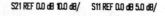

S21 REF 0.0 dB 10.0 dB/   S11 REF 0.0 dB 5.0 dB/

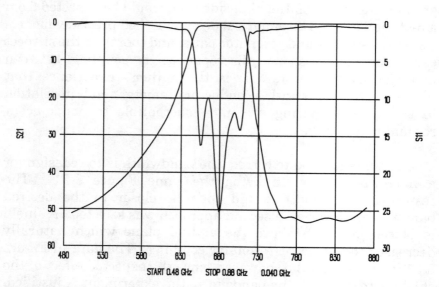

START 0.48 GHz   STOP 0.88 GHz   0.040 GHz

Figure 8-53 *Responses of the wire-over-ground resonator bandpass with a shield over the resonators.*

bandpass filter constructed with 1/16 inch (62 mil) FR4 with resonator unloaded Qs in excess of 500 achieved by a technique which reduces electric fields in the dielectric.

Shown in Figure 8-54 is the two resonator structure used to take the internal coupling data, K. The vertical black rectangular sections are copper strips on each side of the dielectric. They are tied to the same potential by numerous via holes connecting the strips (small circles). The resonator strips are 0.2 inches wide by 1.26 inches long. The horizontal conductor strips are grounded. They rest on a shelf in the aluminum housing for the test board and the final filter. The large diameter via holes clear 4-40 mounting screws. The gray sections in Figure 8-54 are the FR4 dielectric material without copper. The vertical white oval is a section of the FR4 dielectric removed by routing. Housing covers are located 0.25 inches above and below the

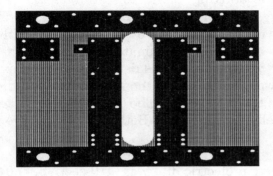

Figure 8-54 *Hidden dielectric resonator pair test board for determination of internal coupling vs. spacing between resonators.*

copper on each side of the PWB. This structure confines electric fields primarily in air and avoids fields in the lossy dielectric material. Fields are near zero in the dielectric between the top and bottom strips because all points along the strips are equipotential, thus the term hidden dielectric. The region between adjacent resonators is devoid of dielectric material because of the routing. The fields are primarily in the air region between the strips and the covers.

Loading capacitors are placed across the gap at the top of the resonators. PWB mounted piston trimmer capacitors are used and are tuned through holes drilled in the side wall adjacent to the trimmers. The large pads on the outside and at the top of the test board are 50 ohm input and output leader lines. Coupling capacitors are placed from these pads to the small pads at the top and side of the resonators. The overall structure is basically combline stripline with an air dielectric, a ground-to-ground spacing of 0.50 inches and a strip thickness of 1/16 inch. This serves as a starting point to estimate the required resonator spacings. Then internal coupling data, K, is taken using procedures outlined in the previous section. Six different test boards with edge-to-edge resonator spacings from 0.20 to 0.40 inches were constructed. This data is presented in Figure 8-54. The data was first taken without removing the dielectric material between the resonators. The data was then taken after routing of the dielectric material. Both sets of data are plotted in Figure 8-55.

The target hidden dielectric filter is a 9-section, 0.0432 dB ripple, Chebyshev bandpass at 1270 MHz with a bandwidth of 64 MHz. The case without the dielectric material removed was chosen to minimize routing costs. The g, k and K values with the resulting spacings are given in Table 8-18. The internal coupling k and K values are computed using techniques described in Section 4.23. The spacings are then read from Figure 8-55.

The required external coupling normalized and unnormalized values are 1.0234 and 20.302 respectively. For this filter,

Table 8-18 *g values, normalized and unnormalized internal coupling data and resonator edge-to-edge spacing for the 9-section 1270 MHz hidden dielectric bandpass filter.*

N	G_n VALUES	$k_{n,n+1}$	$K_{n,n+1}$	$S_{n,n+1}$
1	1.0234	0.8176	0.04121	0.302
2	1.4619	0.5872	0.02960	0.369
3	1.9837	0.5482	0.02763	0.381
4	1.6778	0.5373	0.02708	0.385
5	2.0648	0.5373	0.02708	0.385
6	1.6778	0.5482	0.02763	0.381
7	1.9837	0.5872	0.02960	0.369
8	1.4619	0.8176	0.04121	0.302
9	1.0234			

S21 REF 0.0 dB 10.0 dB/   S11 REF 0.0 dB 5.0 dB/

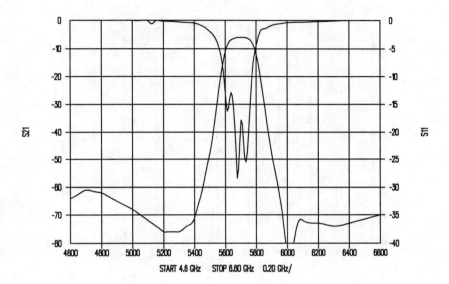

START 4.8 GHz   STOP 6.80 GHz   0.20 GHz/

Figure 8-55 *Internal coupling of the hidden dielectric resonator test board vs. edge-to-edge resonator spacing for notched and unnotched dielectric material.*

external coupling is accomplished using capacitors connected to the top of the input and output resonators. Variable capacitors were used to avoid the need to generate external coupling Q vs. capacitance plots.

Measured responses of the hidden dielectric filter are given in Figure 8-56. A photograph of a similar 9-section 1270 MHz hidden-dielectric bandpass filter with tapped input and output is given in Figure 8-57. In this filter the dielectric material between the resonators is removed by routing.

8.35 Bandpass Tuning Techniques

The predictability and repeatability of PWBs etched from photographic artwork offers the promise of filter manufacture

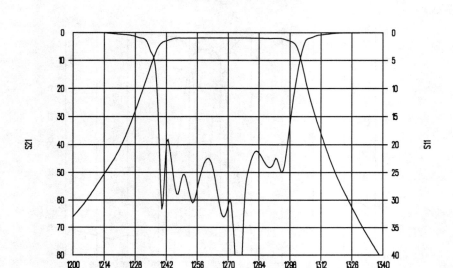

Figure 8-56 *Measurer amplitude transmission and return loss responses of the hidden dielectric bandpass filter.*

without tuning. This goal is readily achieved for lowpass structures except for unusually tight specifications. However, the situation is more severe for the bandpass. For example a Monte Carlo analysis will readily predict component sensitivities increase with decreasing fractional bandwidth. For a 7th order 5% bandwidth 20 dB return loss Chebyshev bandpass, component tolerances of ±0.03% are required for a reasonable degree of confidence that the return loss is not degraded more than 3 dB. This degree of precision from simulation in the design phase is unreasonable. Some adjustment of the design with a prototype is expected for critical applications. Even then, holding the tolerance to ±0.03% for manufacturing runs is a tall order.

The dielectric constant of commercially available materials is normally held to a tolerance of ±2%, with tighter tolerances

Figure 8-57 *Photograph of a 9-section hidden-dielectric 1270 MHz bandpass filter with tapped input and output and routed dielectric.*

more expensive. If the error affects components equally, such as the material dielectric constant, the center frequency is shifted but the return is largely unaffected. A 2% increase in the dielectric constant decreases the center frequency 1%, the center frequency of the example 5% filter is shifted 20% of the bandwidth!

Therefore, as the required bandwidth is decreased, or for tighter specifications, tuning is expected. Fortunately, even when both the inductor and capacitor are inexact, or both the width and length of a resonator are inexact, usually only the center frequency of the resonators must be corrected by tuning. In the 7th order 5% Chebyshev example, the tolerance of the components may be 100 times looser and the 3 dB return loss degradation is still achieved if the resonator center frequencies are corrected.

Which brings us to the problem. How is the filter tuned to obtain the best possible performance? Anyone who has tuned a multi-section filter knows that adjusting all the sections for the best response is like cracking a combination lock. The more sections there are the more difficult it becomes.

Over four decades ago, Dishal [20] published a tuning method which sets all sections close to the required frequency with one pass through the filter. To understand his method, consider Figure 4-2a. When the filter is exactly tuned each branch, whether parallel or series, resonates at f_o. If branch two is removed, the first parallel L-C resonates and a high impedance, limited only by the unloaded Q of the resonator, occurs at the input. If branch three is shorted, the series resonance of branch two causes a low impedance at the input. If branch four is removed, branch three resonates, floating branch two, and again a high impedance occurs at the input. The phenomena repeats and alternate high and low impedances occur at the input resonator at f_o.

A practical procedure results by driving the input with a generator and sampling the voltage across the input resonator with a high impedance probe such as a short wire in the cavity or a very small capacitor connected directly to the input resonator. Ideally, the second resonator is totally removed. In practice, the series capacitor of the series branch is simply shorted. While the latter method is approximate, for narrow bandwidth filters the series inductor, which becomes grounded, is several times larger than the inductor in the shunt resonator and has little effect. The first resonator is tuned for a voltage maximum at f_o at the probe. Next, the third resonator is shorted and the second resonator is tuned for a voltage minimum at f_o. Succeeding resonators are tuned for alternating voltage maximums and minimums.

Bandpass Structures

393

Shown in Figure 8-58 are probed responses in the first resonator of successive resonator tunes of a 5th order 10% bandwidth slabline combline bandpass filter centered at 1000 MHz. Notice the maximum in the plot on the upper left occurs slightly below f_o. The responses in Figure 8-41 are with the resonators tuned exactly as they should be tuned to achieve the optimum response. The Dishal technique calls for the first resonator to be tuned so that the first resonator peaks exactly at f_o. Dishal's technique is an approximation where the quality depends on how effectively a resonator is removed and the percentage bandwidth of the filter. In the illustrated case, the adjacent resonator was shorted (not removed). Notice the ideal resonator tunings do not result exactly with voltage maximums and minimums at f_o. If the resonators are tuned for exact maximums and minimums at f_o, the response on the lower right

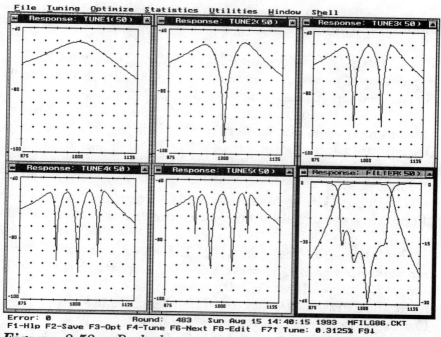

Figure 8-58 *Probed responses at the first resonator for resonators tuned in succession using Dishal's technique. The lower right response results after the first pass.*

in Figure 8-58 results. The return loss worse case is approximately 12 dB and not equal ripple. If the filter is precisely tuned, the resulting return loss for this filter is equal ripple and approximately 18 dB. So while Dishal's technique is not exact, it affords a simple method of tuning the resonators for a first pass. Tuning is completed by fine adjustment of each of the resonators while observing the normal filter responses.

A number of techniques are available to "de-resonate" a section for Dishal's technique. The more effectively a section is removed, the more accurate the first pass will be. Resonators were de-tuned in the slabline combline described in Section 8-16 by utilizing tuning screws which were sufficiently long that when they were fully inserted, they penetrated the hollowed resonator rods until they made contact with the unhollowed section of the rods.

8.36 References

[1] G. Matthaei, L. Young and E.M.T. Jones, *Microwave Filters, Impedance-Matching Networks, and Coupling Structures*, Artech House, Dedham, Massachusetts, 1980, p. 440.
[2] M. Kirschning, R.H. Jansen, and N.H.L. Koster, Measurement and Computer-Aided Modeling of Microstrip Discontinuities by an Improved Resonator Method, *MTT-S Digest*, 1983, p. 495.
[3] ibid., p. 473.
[4] C. Denig, Using Microwave CAD Programs to Analyze Microstrip Interdigital Filters, *Microwave Journal*, March 1989, p. 147.
[5] G. Matthaei, L. Young and E.M.T. Jones, *Microwave Filters, Impedance-Matching Networks, and Coupling Structures*, Artech House, Dedham, Massachusetts, 1980, p. 217.
[6] R.M. Kurzrok, Design of Comb-Line Band-Pass Filters (Correspondence), *MTT-14*, July 1966, p. 351.
[7] G. Matthaei, L. Young and E.M.T. Jones, *Microwave Filters, Impedance-Matching Networks, and Coupling Structures*, Artech House, Dedham, Massachusetts, 1980, p. 497.

[8] W.J. Getsinger, Coupled Rectangular Bars Between Ground Planes, *MTT-10*, January 1962, p. 65.

[9] G. Stracca, G. Macchiarella, and M. Politi, Numerical Analysis of Various Configurations of Slab Lines, *Trans. MTT-34*, March 1986, p. 359.

[10] S. Caspi and J. Adelman, Design of Combline and Interdigital Filters with Tapped-Line Input, *MTT-36*, April 1988, p. 759.

[11] B.C. Wadell, *Transmission Line Design Handbook*, Artech House, Dedham, Massachusetts, 1991, p. 440.

[12] =TLINE= *Operation Manual*, Eagleware Corporation, Stone Mountain, GA, 1992.

[13] M. Dishal, A Simple Design Procedure for Small Percentage Bandwidth Round-Rod Interdigital Filters, *MTT-13*, September 1965, p. 696.

[14] W.A. Davis and P.J. Khan, Coaxial Bandpass Filter Design, *MTT-19*, April 1971, p. 373.

[15] I. Rubenstein, R. Steven and A. Hinte, Narrow Bandwidth Elliptic Function Filters, *MTT-17*, December 1969, p. 1108.

[16] J. Ness and S. Johnson, Narrowband Elliptic Filters on Microstrip, *Microwaves & RF*, November 1984, p. 74.

[17] G.F. Craven and C.K. Mok, The Design of Evanescent Mode Waveguide Bandpass Filters for a Prescribed Insertion Loss Characteristic, *Trans. MTT-19*, March 1971, p. 295.

[18] R.V. Snyder, New Application of Evanescent Mode Waveguide to Filter Design, *Trans. MTT-25*, December 1977, p. 1013.

[19] S.S. Bharj, Evanescent Mode Waveguide to Microstrip Transition, *Microwave Journal*, February 1983, p. 147.

[20] M. Dishal, Alignment and Adjustment of Synchronously Tuned Multiple-Resonant-Circuit Filters, *Elec. Comm.*, June 1952, p. 154.

9

Highpass Structures

This chapter describes a hybrid lumped-distributed highpass filter structure. Highpass filters require series capacitors which are difficult to realize in distributed form. The hybrid highpass uses distributed stubs and series lumped capacitors.

9.1 Overview

The L-C highpass filter structure can be transformed directly from the lowpass prototype values, making this structure as easy to synthesize as the L-C lowpass filter. However, the absence of a convenient series capacitor equivalence presents a realization hurdle for distributed highpass filters. Transmission lines placed end-to-end must have an impractically narrow gap to develop adequate capacitance for highpass filters. Edge-coupled lines increase the adjacent area and therefore the capacitance, but as they become sufficiently long to develop adequate capacitance they no longer behave as pure capacitance. This hurdle is often overcome by using lumped or semi-lumped capacitors. High dielectric constant materials make chip capacitors more practical and inexpensive at microwave frequencies than inductors.

9.2 Stub All-Pole Highpass

Consider a 7th order Chebyshev highpass filter with a cutoff frequency of 2000 MHz. The initial =M/FILTER= screen is shown in Figure 9-1. The substrate is 31 mils thick with a nominal dielectric constant of 2.55 and a loss tangent of 0.001. The impedance of the shorted stubs is 80 ohms. The shorted stubs are the equivalent of shunt inductors in the L-C highpass circuit. In a direct transformation, the series capacitors would be connected directly to the stubs. However, the stubs would then be close enough to couple to each other, degrading filter performance. To solve this problem, line lengths with a specified length (in this case 5 degrees) and a characteristic impedance equal to the input impedance are inserted on each side of the capacitor. To compensate for these lines, their reactance at the cutoff frequency is calculated, and the capacitor is corrected (increased) to make the total reactance at cutoff between the stubs equal to the reactance of the original capacitor at cutoff.

The response of this filter is shown in Figure 9-2. Both the optimized and unoptimized responses are shown. The filter response was close to the 0.25 dB Chebyshev ripple as specified in the =M/FILTER= main window. Optimization has removed much of the ripple. If desired, the ripple could be reinstated and the selectivity improved by modifying the optimization goals and

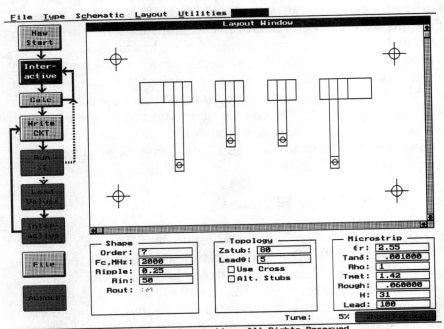

=M/FILTER= (c) 1993 Eagleware Corporation All Rights Reserved

Figure 9-1 *Stub highpass initial =M/FILTER= screen.*

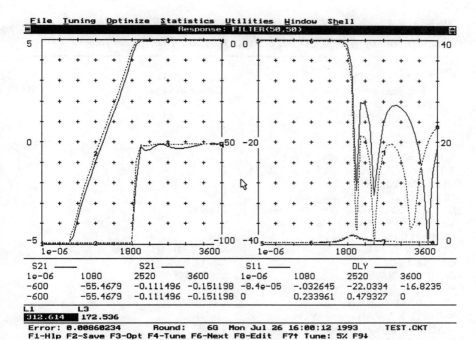

Figure 9-2 *Responses before (solid) and after (dashed) optimization of the 2000 MHz stub highpass microstrip filter.*

reoptimizing. The circuit file written by =M/FILTER= is given in Table 9-1.

Distributed filter structures are typically designed with lines which are operated at or below the first resonant frequency. Reentrance therefore occurs higher than the passband frequency. For lowpass and bandpass structures, reentrance therefore occurs in the stopband. Stopband attenuation is degraded but the passbands are not compromised. For highpass filters, reentrance occurs above the cutoff within the passband. Therefore, distributed highpass filters have a limited passband bandwidth. They are to a certain degree bandpass structures.

Figure 9-3 shows the very high frequency response of this filter. Note that even though this filter is called a highpass, it actually does have stopbands above cutoff. The first reentrance stopband

Table 9-1 *Circuit file written by =M / FILTER= with a physical description of the 2000 MHz hybrid stub highpass filter.*

```
'   FILE: MFILF91.CKT              CL2=41.3784
'   TYPE: Stub -- Highpass         CL3=41.3784
'     Fc: 2000 MHz                 CL4=41.3784
' PROCESS: Microstrip              CL5=41.3784
                                   CL6=41.3784
CIRCUIT                            CL7=41.3784
SUB ER=2.55 TAND=0.001 RHO=1       CL8=41.3784
& TMet=1.42 ROUGH=0.06 UNITS=0.0254  CLW=84.9593
MLI 1 2 W=WI H=H L=LI              WI=84.9593
MLI 2 3 W=CLW H=H L=CL1            LI=100
MTE 3 4 100 WT=CLW WS=W1 H=H       L1=?262.534
MLI 100 101 W=W1 H=H L=L1          W1=37.0262
MVH 101 0 R=VIAR H=H T=VIAT        VIAR=20
MLI 4 5 W=CLW H=H L=CL2            VIAT=0.71
CAP 5 6 C=C2                       C2=1.04004
MLI 6 7 W=CLW H=H L=CL3            L3=?164.369
MTE 7 8 110 WT=CLW WS=W3 H=H       W3=37.0262
MLI 110 111 W=W3 H=H L=L3          C4=0.968563
MVH 111 0 R=VIAR H=H T=VIAT        L5=L3
MLI 8 9 W=CLW H=H L=CL4            W5=37.0262
CAP 9 10 C=C4                      C6=1.04004
MLI 10 11 W=CLW H=H L=CL5          L7=L1
MTE 11 12 120 WT=CLW WS=W5 H=H     W7=37.0262
MLI 120 121 W=W5 H=H L=L5          WOUT=84.9593
MVH 121 0 R=VIAR H=H T=VIAT        LOUT=100
MLI 12 13 W=CLW H=H L=CL6          WINDOW
CAP 13 14 C=C6                     FILTER(50,50)
MLI 14 15 W=CLW H=H L=CL7          GPH S21 -5 5
MTE 15 16 130 WT=CLW WS=W7 H=H     GPH S21 -100 0
MLI 130 131 W=W7 H=H L=L7          GPH S11 -40 0
MVH 131 0 R=VIAR H=H T=VIAT        GPH DLY 0 40
MLI 16 17 W=CLW H=H L=CL8          FREQ
MLI 17 18 W=WOUT H=H L=LOUT        SWP 0 3600 41
DEF2P 1 18 FILTER                  OPT
EQUATE                             2000 3600 S11<-100
H=31                               400 1200 S21<-30
CL1=41.3784
```

is above 10 GHz, five times higher than the design cutoff frequency. Shorter stubs increase the reentrance frequency and

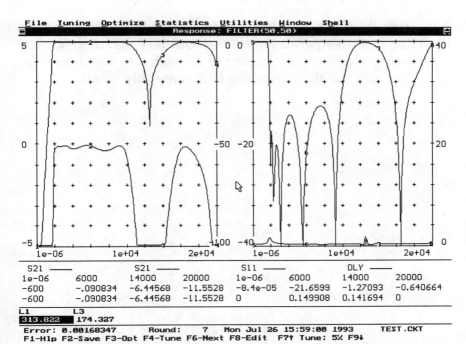

Error: 0.00168347   Round: 7   Mon Jul 26 15:59:00 1993   TEST.CKT
F1-Hlp F2-Save F3-Opt F4-Tune F6-Next F8-Edit F7↑ Tune: 5% F9↓

Figure 9-3 *Wideband sweep of the stub highpass filter showing reentrance in the passband above approximately 10 GHz.*

widen the useful passband of the highpass. Reentrance can be pushed as high as required except that shorter stubs must be of higher impedance to achieve the required inductance.

Two additional options are available in the stub highpass. The first option is to use crosses and double stubs. If this option is used, for a given specified Zstub, the line lengths are increased and the passband bandwidth is reduced. Therefore the double stub option is most useful when high line impedances are possible.

The final option is to alternate stubs which directs the first stub down, the next stub up, and so on. If stubs are alternated, the Lead θ, which is 5 degrees in this example, can generally be reduced since the stubs will be much farther apart and coupling will not be a problem. This reduces the overall filter length, but nearly doubles the width because the stubs extend in both directions. Filter performance is essentially unchanged. The alternating stubs option is especially useful as the frequency increases and the filter becomes unmanageably small.

10

Bandstop Structures

When the rejection of a signal is required, it is natural to think in terms of a notch. Just as a true bandpass filter offers improved selectivity over a single resonator, the bandstop filter offers improved rejection over a simple notch or even a cascade of notches. However, the general realization difficulties of distributed structures are worsened by particular difficulties associated with the bandstop structure.

10.1 Overview

The distributed bandstop structures can theoretically be transformed directly from a conventional L-C bandstop filter. However, in practice, this direct transformation gives unrealistic line impedances, often on the order of hundreds of ohms. Kuroda's identities can be used to help this process, inserting admittance inverters and forcing the line impedances to become somewhat more reasonable. An edge coupled bandstop filter is also available, but is only realizable for very narrow bandwidths.

10.2 Stub Bandstop

An exact set of design equations is available for 1st to 5th order stub bandstop filters [1]. These equations give impedances for both cascade and stub lines. All lines are 90 degrees long. While there is no mathematical limit to the bandwidth of this filter, the stub impedances quickly become impractical as the bandwidth decreases below about 50 percent for printed structures or about 30 percent for mechanical structures. If a narrower bandwidth is required, the open-circuited stub resonators can be replaced with capacitively coupled short-circuit resonators.

The synthesis equations are based on Kuroda's identity. This identity allows shifting stubs and conversion of a series stub to a shunt stub while making the impedance levels more realizable. The following design equations can be derived for 3rd order filters.

$$Z_1 = Z_A \left(1 + \frac{1}{\Lambda\ g_0\ g_1} \right) \tag{1}$$

$$Z_{12} = Z_A (1 + \Lambda\ g_0\ g_1) \tag{2}$$

$$Z_2 = \frac{Z_A\ g_0}{\Lambda\ g_2} \tag{3}$$

$$Z_3 = \frac{Z_A\ g_0}{g_4} \left(1 + \frac{1}{\Lambda\ g_3\ g_4} \right) \tag{4}$$

$$Z_{23} = \frac{Z_A\ g_0}{g_4} (1 + \Lambda\ g_3\ g_4) \tag{5}$$

$$\Lambda = \cot\left(\frac{\pi\,\omega_1}{2\,\omega_0}\right) \tag{6}$$

$$Z_B = \frac{Z_A\,g_0}{g_4} \tag{7}$$

$$\omega_o = \frac{\omega_1 + \omega_2}{2} \tag{8}$$

where Z_A is the input impedance of the filter, Z_B is the output impedance of the filter, ω_1 is the lower passband corner radian frequency and ω_2 is the upper passband corner. While these design expressions are exact for wide and narrow bandwidth, realizability in the stub form is only practical for wide bandwidth. An extension of this process is used by =M/FILTER= to synthesize stub bandstop filters to 21st order.

Consider a 3rd order Chebyshev bandstop filter with a stopband of 2000 to 3500 MHz. The initial =M/FILTER= screen is shown in Figure 10-1. The substrate is 31 mils thick with a nominal dielectric constant of 2.55 and a loss tangent of .001. This example illustrates the realization difficulties of this structure. Even with this 55% bandwidth, the end stub lines are only 9 mils wide. As the filter bandwidth gets narrower, the stub width decreases even further, making this filter impractical for narrow or even moderate bandwidth on printed boards.

The response of this filter is shown in Figure 10-2. This response did not need optimization in =SuperStar=. The circuit file is shown in Table 10-1. All lines in this filter are 90 degrees long at the center frequency and the line impedances are determined automatically. The two topology options available to the designer are to use crosses and to alternate the stubs. If

Use Crosses is selected, any line whose impedance goes below Zswi gets converted into a double stub with each line having twice the original impedance of the original and a length of 90 degrees. Because the primary realization difficulty with this filter is high stub line impedance, crosses worsen realizability and should be avoided except for very wide bandwidth.

The alternate stubs option is useful for higher frequencies where the stubs approach each other and couple. When crosses are not necessary (or selected) and this option is used, the lines alternate pointing down and up. The circuit file used for the physical implementation of the filter in Figure 10-1 is shown in Table 10-1.

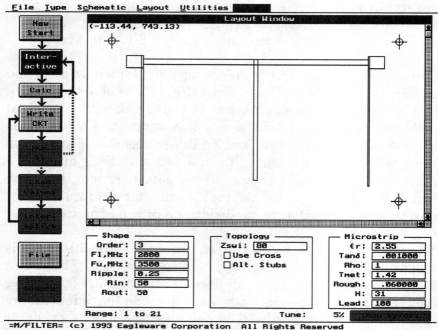

Figure 10-1 *Stub bandstop initial =M/FILTER= screen.*

10.3 Edge-Coupled Bandstop

The previous stub bandstop is suitable for wide bandwidth applications. For narrow bandwidth, the edge-coupled bandstop is useful. Various configurations of edge-coupled bandstop filters are available. The configuration discussed here has all 90 degree resonator sections separated by lines of 90 degrees. The design equations for this filter are also based on Kuroda's identities and are very similar to the equations for the stub bandstop filter. Reference [2] gives exact equations for 1st to 5th order filters. Although the synthesis equations are exact for wide and narrow bandwidth, realization is practical only for narrow to moderate bandwidth in mechanical form and narrow bandwidth in microstrip. =M/FILTER= uses an extension of this synthesis procedure to design bandstop filters to 21st order.

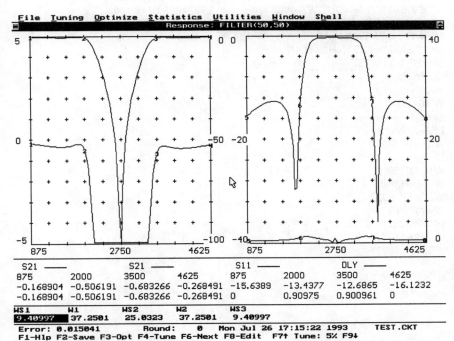

Figure 10-2 *=SuperStar= response of a stub bandstop filter designed by =M/FILTER=.*

Table 10-1 Circuit file for edge coupled bandstop filter implemented in slabline.

```
'    FILE: MFILF101.CKT
'    TYPE: Stub -- Bandstop
'     FI: 2000 MHz
'     Fu: 3500 MHz
' PROCESS: Microstrip

CIRCUIT
SUB ER=2.55 TAND=0.001 RHO=1              WINDOW
& TMet=1.42 ROUGH=0.06 UNITS=0.0254      FILTER(50,50)
MLI 1 2 W=WI H=H L=LI                     GPH S21 -5 5
MTE 2 3 100 WT=WI WS=WS1 H=H             GPH S21 -100 0
MLI 100 101 W=WS1 H=H L=LS1             GPH S11 -40 0
MEN 101 0 W=WS1 H=H                       GPH DLY 0 40
MLI 3 4 W=W1 H=H L=L1                     FREQ
MTE 5 4 110 WT=W2 WS=WS2 H=H            SWP 875 4625 101
MLI 110 111 W=WS2 H=H L=LS2            OPT
MEN 111 0 W=WS2 H=H                       2375 3125 S21<-30
MLI 5 6 W=W2 H=H L=L2                     875 2000 S11<-100
MTE 7 6 120 WT=WOUT WS=WS3 H=H          3500 4625 S11<-100
MLI 120 121 W=WS3 H=H L=LS3
MEN 121 0 W=WS3 H=H
MLI 7 8 W=WOUT H=H L=LOUT
DEF2P 1 8 FILTER
EQUATE
H=31
WI=84.961
LI=100
LS1=780.872
WS1=?9.40997
L1=747.785
W1=?37.2501
LS2=766.678
WS2=?25.0323
L2=747.785
W2=?37.2501
LS3=780.872
WS3=?9.40997
L3=0
WOUT=84.961
LOUT=100
```

Figure 10-3 shows the initial =M/FILTER= screen for a 3rd order edge-coupled bandstop filter with a stopband from 2000 to 2400 MHz (18% bandwidth) implemented in slabline with air dielectric. The ground to ground plane spacing is 500 mils. The edge-coupled bandstop is generally much more lossy than the stub bandstop filter and is best implemented mechanically or with thick substrates using low-loss dielectric materials. Implementation in microstrip is generally poor due to loss for narrow bandwidths, and the tight spacings required for moderate to wide bandwidths. The resonators used in this example are all 50 ohms. This filter type generally requires significant optimization of the initial values. The number of coupled sections is equal to the order with each coupled section being 90 degrees long. The lines separating the coupled sections are also each 90 degrees long. Because the electrical length is

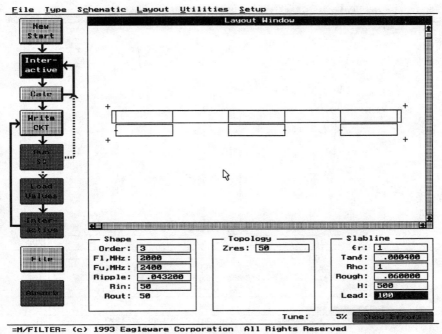

=M/FILTER= (c) 1993 Eagleware Corporation All Rights Reserved

Figure 10-3 *Initial =M/FILTER= screen for an edge-coupled 2200 MHz bandstop filter for slabline.*

180 degrees per section (less 90 degrees), this structure tends to be physically long at lower frequencies.

The only option available to the designer is the choice of impedance level for the transmission lines used in the filter. A higher impedance level will generally increase the required spacings, making a wider bandwidth filter more realizable. Due to coupling effects, the line widths required to keep the resonator impedances identical may vary. If this is not desired, all line widths may be set equal to each other before optimization.

The filter response is shown in Figure 10-4. Note that the filter required significant optimization to obtain the expected response. Prior to optimization, the passband ripple was over 2.5 dB and the return loss had 4 dB peaks just below the lower

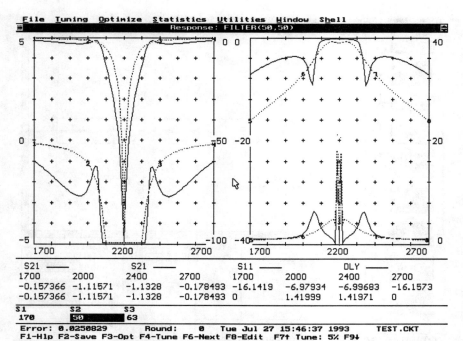

Figure 10-4 *Response of the slabline edge coupled bandstop filter.*

cutoff and just above the upper cutoff. Note the loss of about 1 dB in the passband even when this structure is implemented at a relatively wide 20% bandwidth using low-loss slabline.

The circuit file used for this filter prior to optimization is shown in Table 10-2. RLI is the circuit file code for round-rod lines between ground planes RCP is the code for coupled round-rods. A rod diameter of approximately 274 mils with a ground-to-ground spacing of 500 mils results in a 50 ohm line impedance. The total length for this 2200 MHz three-section bandstop is 6.7 inches not including the input and output leader lines. After optimization the rod spacings are 170, 50 and 63 mils.

The group delay of the edge-coupled bandstop is given on the left in Figure 10-4. The group delay is the composite of the behavior

Table 10-2 *Circuit file for slabline edge coupled bandstop filter.*

' FILE: TEST.CKT	D1=274.393
' TYPE: Edge Coupled -- Bandstop	S1=?41.9515
' Fl: 2000 MHz	L1=1341.22
' Fu: 2400 MHz	D2=274.393
' PROCESS: Slabline	S2=?41.9515
	L2=1341.22
CIRCUIT	D3=274.393
SUB ER=1 TAND=0.0004 RHO=1	S3=?41.9515
& TMet=0.71 ROUGH=0.06 UNITS=0.0254	L3=1341.22
RLI 1 2 D=DI H=H L=LI	DOUT=274.393
RCP 2 0 98 3 D=D1 S=S1 H=H L=L1	LOUT=100
RLI 3 4 D=WR H=H L=LZ	WINDOW
RCP 4 107 0 5 D=D2 S=S2 H=H L=L2	FILTER(50,50)
RLI 5 6 D=WR H=H L=LZ	GPH S21 -5 5
RCP 6 0 116 7 D=D3 S=S3 H=H L=L3	GPH S21 -100 0
RLI 7 8 D=DOUT H=H L=LOUT	GPH S11 -40 0
DEF2P 1 8 FILTER	GPH DLY 0 40
EQUATE	FREQ
H=500	SWP 1700 2700 101
DI=274.393	OPT
LI=100	2100 2300 S21<-30
WR=274.393	1700 2000 S11<-100
LZ=1341.23	2400 2700 S11<-100

of lowpass and highpass filters. Group-delay equalization in the passband below the lower cutoff is achieved using techniques identical to lowpass filter equalization. As with a highpass filter, the upper passband extends to infinite frequency and group-delay equalization of the entire passband requires an infinite number of all-pass sections. The transmission phase is discontinuous at the center frequency transmission zeros and therefore the group-delay is undefined.

10.4 References

[1] G. Matthaei, L. Young and E.M.T. Jones, *Microwave Filters, Impedance-Matching Networks, and Coupling Structures*, Artech House Books, Norwood, Massachusetts, 1980, p. 757.

[2] Dipak S. Kothari, Exact Solution for a Multi-Section Stopband Filter with Resonators 90 degrees Apart, *Microwave Journal*, Norwood, Massachusetts, March 1988, p. 183.

Appendix A

PWB Manufacturing

PWB manufacturing requirements can be divided into two classes: prototype and production quantities. Prototype boards have traditionally been constructed using a photographic etching process, although milling is becoming popular for single-layer prototypes due to the fast turn-around time and reduced setup costs. Production quantities are seldom milled due to a large processing time per board and are more often etched or deposited.

A.1 Photographic Etching

The photographic etching process is efficient for manufacturing large quantities of circuit boards, having been used now for several decades. Etching involves several steps. The first is the generation of artwork on transparent film. Various methods can be used by the designer to generate film, including RubyLithe, photoplotting, linotype, and computer printer transparencies. Alternatively, service bureaus are available which take a desktop CAD/publishing file which the designer creates (such as DXF or Gerber) and generate film using high precision equipment. The board manufacturer's responsibility is to replicate the film pattern as metalization on the desired substrate material. The manufacturer starts with a metalized board, drills via holes, replates the entire board to plate the via holes, applies photoresist to the board, exposes the photoresist to ultraviolet light using the transparent film, and chemically etches the unwanted metal from the board. Each of these steps brings with it additional factors which must be considered for accurate microwave filter design.

An important consideration in the etching process is the "etch factor." This is the amount by which the chemical etching process over-etches the board, as shown in Figure A-1. This etch factor is accounted for in the film generation step to avoid errors in line widths and lengths. The etch factor varies among manufacturers and should be discussed with the board house on a case by case basis. The etch factor correction is implemented as a thickening of the traces on a circuit. Figure A-2 shows a circuit before and after the etch factor is added. The etch factor used here is exaggerated to clarify the effect. A typical etch factor is 1 mil. Note that accurate etch factor compensation is not as simple as changing the widths of each line by the etch factor because reference planes are also shifted. Compensation is best done with the aid of software.

A second consideration in any process which adds metal to a circuit board (such as plating for via holes or tin-lead final plating) is the metal thickness of the signal layer. (The thickness of the ground plane is not overly important provided it is thicker than the skin depth.) The metalization thickness must be taken into account at the very start of the design because it affects the characteristic impedance and therefore the width of the lines. Figure A-3 shows the required line widths for varying impedances and metallization thicknesses on PTFE (ε_r=2.2) with a 22 mil (0.559 mm) dielectric thickness. Note that the line

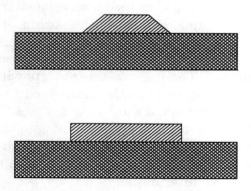

Figure A-1 *Etched board (top) compared to ideal board (bottom). After etching the line width is narrower and the edges are sloped.*

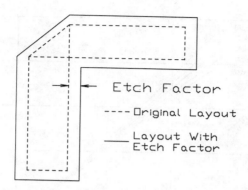

Etch Factor

---- Original Layout

____ Layout With
Etch Factor

Figure A-2 *Etch factor for a 90 degree bend. Notice that both the line width and the corner reference planes are affected.*

widths can be off by as much as 15 percent for high impedance lines, which could have a severe impact on filter performance. The metalization thickness also impacts coupled line performance. A metalization thickness of zero represents the ideal thin-line case and $t=2.42$ mils represents 2 ounce copper.

After the circuit is designed, the next step is creating the artwork. Artwork is generated manually using RubyLithe and a hand cutting tool or using pen, paper, and a camera. Manual artwork generation is inexact, and human error often overshadows other considerations such as etch factor and metalization thickness. The advent of personal computers has popularized automated artwork generation. The engineer may choose between a number of software packages specifically created for layout generation or art design. Output is sent to a printer, plotter, or outside service bureau for film generation.

Laser printers are used to output artwork for less critical applications. They are affordable and readily available, but are less accurate than high-quality pen and photoplotters. For improved accuracy, laser printers are manually "calibrated" by printing a solid square of known dimensions, the larger the better. The printer is set dark enough to insure that the square

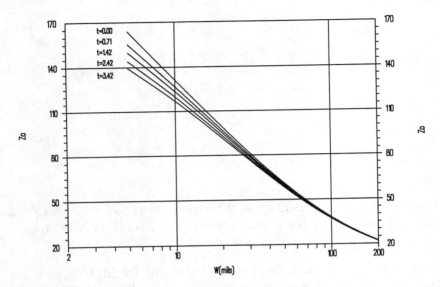

Figure A-3 *Required line widths vs. impedances and metalization thicknesses in mils on PTFE (ε_r=2.2) with a 22 mil (0.559 mm) dielectric thickness.*

is solid. Next, the actual horizontal and vertical size of the square is measured using an accurate measuring device such as calipers and a measuring microscope platform. Dividing the expected size by the actual size gives horizontal and vertical scaling factors. The artwork should be scaled in the software by these scaling factors. It is advisable to save an original, unscaled version of the artwork file to facilitate later rescaling. The calibration is dependent on the paper or transparency type and generally varies with time and temperature.

A second and more difficult to measure factor is the "bleed" on sharp edges. The behavior of bleeding is similar to the etch factor described earlier except that lines get wider instead of narrower. This bleeding can be corrected with a negative etch factor. To measure the bleeding factor, several sample squares (at least two) of different sizes are created and measured. The following formula is then solved for both the horizontal and vertical direction.

$$ActualSize = ExpectedSize \div ScaleFactor + Bleed$$

A second source of error in laser printing of film is acetate temperature instability. To minimize the temperature effect, acetates should be at a consistent temperature before printing and during board exposure. A third cause of error is laser printer resolution. A 300 dpi resolution laser printer has a potential error of ± 1.7 mils ($\pm .04$ mm). Newer laser printers with resolutions of 600, 800, or even 1200 dpi minimize this error. Laser printers which use microfine toner generally produce blacker output which is less likely to require manual touch-up. Resolution, scaling, and bleeding errors are reduced printing the artwork with as large a scale as possible provided photographic reduction of the artwork is possible.

Pen plotters are used to generate film, paper, or RubyLithe artwork. Pen plotters generally have better resolution than laser printers (1016 dpi) but have the disadvantage of much larger bleed due to finite pen width. Pen plotters are also loaded with knife blades to cut RubyLithe. RubyLithe is a two layer sheet stuck together with weak adhesive. The bottom layer is transparent, and the top layer is opaque and ruby-colored. When cutting RubyLithe, the plotter draws the outlines of the object. Most CAD software packages output to HPGL compatible pen plotters.

The most accurate method to generate film is by photoplotter or linotype. Films are often generated by an outside service bureau due to the high cost of the output devices. Most photoplotters accept Gerber format data, while most linotypes accept PostScript format. Gerber data is more often available in CAD packages because the Gerber format was designed for CAD while PostScript was designed for page layout. A number of drawing packages exist for converting HPGL or AutoCAD DXF files to Gerber format and for Gerber graphic editing.

Since photoplotting is generally handled by a service bureau, it is important to insure the correct output file format which varies among service bureaus. For example, Gerber devices vary in the supported command set and decimal format used. Output files are generally sent to the bureau on disk or tape, but for faster turnaround time, many services will accept files via modem (digital transfer via phone). Some PWB manufacturers also have photoplotting equipment available. One bureau which was used in generating film for test filters studied in this book is:

Advance Reproductions Corporation
100 Flagship Drive
North Andover, MA 01845
Phone (508) 685-2911 FAX (508) 685-1771
Contact: Steve Alaimo or Leslie Townsend
Gerber format data preferred; HPGL or DXF also accepted
Send data via 3.5" or 5.25" disks, 9 track mag tape, or modem
Notes: Provider of high precision CAD and photolithography
       services to hybrid, microwave, semiconductor, and PCB
       manufacturers. Overnight photoplotting guaranteed.

After film is generated, it is used by the PWB manufacturer to etch the boards. Three manufacturers created test boards for this book. They are:

Microwave Printed Circuitry, Inc. (MPC)
81 Old Ferry Road
Lowell, MA 01854
Phone (508) 452-9061 FAX (508) 441-0004
Technical Contact: Christopher Bradford, Engineering Manager
Purchasing Contact: Bob Deitz, National Sales Manager
Preferred Formats: DXF or camera-ready film
Stocked PWB Material: Selected stock of Arlon, Rogers,
       Taconics, and Norplex/Oak
Note: MPC specializes in PTFE based printed wiring
       boards for microwave frequency applications.

Lehighton Electronics
First and South Street
Lehighton, PA 18235
Phone (215) 377-5990 FAX (215) 377-5990
Contact: Norma Felter
Preferred Format: Camera-ready film
Stocked PWB Material: Various PTFE

For the test filters, DXF files were generated. These files were then converted to Gerber format using software supplied by:

CAD Design Services, Inc.
2806-C Soquel Avenue
Santa Cruz, CA 95062
Phone (408) 462-6494 FAX (408) 475-0738
Contact: John Sovinsky

A.2 Machining

Machining is a process by which the unwanted copper is machined, or milled, from the circuit board. Automated machining offers exceptionally short design to test times. The milling process requires very little setup, but the time per board after setup is large. For example, it may take thirty minutes to convert an artwork file and setup the milling machine and each board might take twenty minutes to mill. In etching, the same board may take hours or days to set up but manufacture of production quantities is more efficient. Therefore machining is generally used for prototypes rather than production.

The resolution of milling machines is typically 0.25 mils (0.006 mm). The minimum trace width is 4 mils (0.1 mm) and the minimum spacing between lines or other features is 8 mils (0.2 mm). The minimum spacing is relatively large due to the need to get a milling bit into the space. Machined circuit boards will have traces with square edges, as compared to the somewhat beveled edges characteristic of chemically etched boards. Machined circuit boards generally also have about 1 mil (0.025

mm) of board material stripped off. These differences are only significant for thin boards or closely spaced lines.

Many filters described in this book were milled using the Quick Circuit 7000 machine shown in Figure A-4 which is produced by:

T-Tech, Inc.
5591-B New Peachtree Road
Atlanta, GA 30341
Phone (404) 455-0676
FAX (404) 455-0970

DXF files generated by =M/FILTER= were read by GerbArt software from CAD Design Services (address given above) which created Gerber format files. Quick Circuit software and hardware were then used to produce the prototypes.

Figure A-4 *Quick Circuit 7000 PWB milling machine from T-Tech, Inc.*

Appendix B

Variable Definition Lists

This appendix defines equation variables (symbols) used throughout the book. These variables are listed in this appendix in alphabetical order. This appendix also briefly describes =SuperStar= circuit file model codes and variables written by the =M/FILTER= program to define physical dimensions of distributed filter structures. Model codes are organized by process and by function. Circuit file dimensional variables are listed in alphabetical order.

B.1   Equation Variables

When possible, equation variables (symbols) are defined consistently throughout the book. In certain cases, the original contributor's definitions are retained to avoid reformulation of the original equations. This introduces some unavoidable inconsistencies. In general, definitions are obvious if a chapter is read sequentially. Temporary variables are defined for local use only.

A	amplitude response
A_{min}	minimum attenuation
a	radius of the inner conductor of a coaxial line
a_t	line attenuation
a_1	incident S-parameter voltage wave at the network input divided by the square root of the reference impedance
a_2	incident S-parameter voltage wave at the network output divided by the square root of the reference impedance
B	magnetic vector field
$B_{n,n+1}$	susceptance of series coupling capacitors between resonator n and n+1

b	radius to the inside of the outer conductor of coax
b_1	reflected S-parameter voltage wave at the network input divided by the square root of the reference impedance
b_2	incident S-parameter voltage wave at the network input divided by the square root of the reference impedance
C_t	total normalized node capacitance
C_{11}	linear S-parameter input reflection coefficient
C_{21}	linear S-parameter forward transmission coefficient
C_{12}	linear S-parameter reverse transmission coefficient
C_{22}	linear S-parameter network output reflection coefficient
C	capacitance in Farads (1F = 1 Coulomb/Volt)
C_{hp}	highpass capacitor value
C_{lp}	lowpass capacitor value
C_e	effective capacitance of an inductor
c	speed of light (2.9979246×10^8 m/s)
d	diameter of a single conductor transmission line
E	potential difference between nodes
E_{avail}	load voltage when all available generator power is delivered to the load
E_g	voltage of the generator (source)
E_l	voltage across the load
f_c	cutoff frequency of a low or highpass process
f_l	lower cutoff frequency of a bandpass or bandstop
f_o	center frequency of bandpass or bandstop process
f_r	resonant frequency
f_u	upper cutoff frequency of a bandpass or bandstop
$g(0)$	normalized input termination resistance for lowpass prototypes
$g(N+1)$	normalized output termination resistance for lowpass prototypes of order N
$g(n)$	nth g-value for a lowpass prototype
H	1) voltage attenuation coefficient (H(s) is the attenuation transfer function), 2) Medhurst's effective capacitance factor, 3) magnetic field intensity or magnetizing force
I	conductor current
j	square root of -1 (i is also commonly used)
$J_{n,n+1}$	admittance inverter parameters

$K(k)$	complete elliptic integral of the first kind of modulus k
$K'(k)$	complete elliptic integral of the first kind of complementary modulus k', $K'(k)=K(k')$
$K(s)$	characteristic function of the Feldtkeller equation
k	transmission line coupling coefficient
k_c	constant factor for adjusting the poles of a Butterworth polynomial to create a Chebyshev response
L	inductance in Henries ($1H = 1$ Volt·sec^2/Coulomb)
L_A	insertion loss resulting from reflected energy
L_{Ar}	passband attenuation ripple
L_{hp}	high pass inductor value
L_{lp}	low pass inductor value
L_t	total resonator length
L_{tu}	total resonator length relative to a quarter-wavelength
N	Filter order, or number of turns in an inductor
N_f	demagnetization factor
n	index variable used in numerous equations
P_i	the power incident
P_l	power delivered to load with the network present
P_{null}	power delivered to the load with a null network
P_r	the radiated power
Q_d	Q resulting from dielectric loss
Q_e	effective unloaded Q
Q_u	unloaded Q of a purely reactive component (L or C)
Q_{uo}	unloaded Q without parasitic capacitance
RL	return loss, decibel format, positive for passive networks
R_{ac}	high frequency resistance of a conductor
R_{dc}	resistance of a conductor at 0 Hz (dc)
R_g	resistance of the generator
R_l	output termination resistance (load)
R_p	parallel loss resistance
R_s	series loss resistance
r	solenoid radius in inches to the wire center
S	spacing between capacitor plates
S_r	hairpin resonator spacing
S_{11}	S-parameter network input reflection coefficient
S_{21}	S-parameter forward transmission coefficient

424

S_{12}	S-parameter reverse transmission coefficient
S_{22}	S-parameter network output reflection coefficient
s	complex phasor variable ($s=\sigma + j\omega$)
	also center to center wire spacing of a solenoid
t	voltage transmission coefficient, or toroid thickness
t_d	group delay of a network
t_p	phase delay of a network
$VSWR$	voltage standing wave ratio
W,w	width of a metalization strip
X	reactance of a component, or impedance j coefficient
Y_o	characteristic admittance = $1/Z_o$
$y_{n,n+1}$	admittances of the series transmission lines in the equivalent wire-line model of combline
Z	terminal impedance
Z_A	input impedance of a filter
Z_B	output impedance of a filter
Z_i	input impedance (also Z_{input})
Z_m	arithmetic mean of the even and odd mode impedances
Z_o	characteristic impedance of a transmission system
Z_{oe}	even mode line impedance of coupled lines
Z_{oo}	odd mode line impedance of coupled lines
α_c	attenuation from conductor loss
α_d	attenuation from dielectric loss
Δ_{rms}	root mean square of surface roughness, assuming periodic ridges transverse to the current flow
δ	skin depth of a conductor
ε	dielectric constant given by $\varepsilon=\varepsilon_r\varepsilon_o$
ε_o	permittivity of free space (8.8541843 pF/cm)
ε_r	relative dielectric permittivity
η	intrinsic impedance of free space
η_o	characteristic impedance of free space ($\eta_o=120\pi$ Ω, $\varepsilon_r=\mu_r=1$)
θ	modular angle for a Cauer-Chebyshev process
θ_n	electrical length of a resonator (in degrees)
λ	wavelength of input form
λ_{go}	wavelength while traveling in a conductor
λ_o	wavelength while traveling in air
μ	conductor permeability
μ_e	effective permeability of a substance
μ_o	permeability of free space ($\mu_o=4\pi$ nH/cm)

μ_r	permeability of a conductor relative to μ_o
ρ	voltage reflection coefficient
ρ_{eff}	effective resistivity increased by surface roughness
ρ_i	input voltage reflection coefficient
ρ_r	resistivity of a conductor, relative to copper
σ	1) real portion of a complex phasor, or 2) conductivity of a substance ($\sigma=1/\rho_r$)
Φ	electrical length from resonator ground to the tapped point on an interdigital bandpass filter
ϕ	phase angle offset in a complex transfer function
ψ	Medhurst factor in the solenoid unloaded Q expression
ψ_{opt}	Medhurst factor for optimum unloaded Q
ω	angular frequency (usually given in rad/s)
ω_n	frequency normalized by the cutoff frequency
ω_s	cutoff-normalized lowest stopband frequency at which A_{min} occurs

B.2   Circuit File Codes

The following is a short-form list of =SuperStar= Professional element codes. The left-most portion specifies the format of the circuit block line for each code. The right-most portion gives a description of the element.

EXACT MODELS

RES	resistor
IND	inductor
CAP	capacitor
PLC	parallel inductor capacitor
PFC	parallel resonator, frequency & capacitor
PFL	parallel resonator, frequency & inductor
PRC	parallel resistor and capacitor network
PRL	parallel resistor and inductor network
PRX	parallel resistor inductor capacitor
SFC	series resonator, frequency & capacitor
SFL	series resonator, frequency & inductor
SLC	series inductor capacitor
SRC	series resistor and capacitor network
SRL	series resistor and inductor network

SRX series resistor inductor capacitor
CPL coupled lines, electrical parameters
CPNx electrically multiple coupled lines
GYR gyrator
MUI mutually coupled inductors
TRF ideal transformer
TLE electrical transmission line
TLE4 four terminal electrical line
TLP physical transmission line
TLP4 four terminal physical line

DEVICES
BIP bipolar transistor
FET FET transistor
OPA operational amplifier
TWO read S- or Y-parameters for a device
VCC voltage controlled current source

PHYSICAL LINE MODELS and DISCONTINUITIES
CEN coaxial end
CGA coaxial gap
CLI coaxial line
CLI4 four terminal coaxial line
CST coaxial step
MBN microstrip bend
MCP microstrip coupled line
MCNx multiple coupled microstrip lines
MCR microstrip cross
MEN microstrip end
MGA microstrip gap
MLI microstrip line
MRS microstrip radial stub
MST microstrip step
MTE microstrip tee junction
MVH microstrip via hole to ground
RCNx multiple coupled slablines
RCP coupled slabline
RLI slabline (rod between ground planes)
SBN stripline bend
SCNx multiple coupled striplines
SCP stripline coupled line

SEN	stripline end
SGA	stripline gap
SLI	stripline
SSP	stripline step
STE	stripline tee junction
SUB	substrate description for physical models
WAD	waveguide adapter
WLI	rectangular waveguide
XTL	piezoelectric resonator

B.3 Program and Circuit File Variables

The following variables are used in =SuperStar= Professional circuit files written by the =M/FILTER= program.

AS	option specifier for asymmetric step
ER	relative dielectric constant
CEND	capacitance for the end resonators of combline
CH	option specifier for chamfered corner bend
CLn	length of capacitor pad-lines in hybrid highpass
CLW	width of capacitor pad-lines in hybrid highpass
CMID	capacitance for the internal resonators of combline
Cn	capacitance for the nth resonator in combline
D	diameter of a round-rod line
DI	diameter of an input round-rod leader line
DOUT	diameter of an output round-rod leader line
Fc	cutoff frequency of lowpass or highpass filters
Gn	gaps between the ends of lines (n is an integer)
H	microstrip dielectric thickness ("b" for stripline)
LA	length of interdigital end-resonator compensating line
LI	length on an input leader line
LMATCHn	length of matching lines for elliptic bandpass
Ln	length of individual lines (n is an integer)
LOUT	length of an output leader line
LRn	length of resonator lines (stubs) in elliptic bandpass
LSAn	length of a slide segment in edge-coupled and hairpin
LSBn	see LSAn above. n is an integer
LTAn	length of a tapped segment in edge-coupled and hairpin
LTBn	see LTAn above. A at input, B at output, n=1 or 2

LZ	length of uncoupled cascade segments in edge-coupled Bandstop
NAR	width on one side of a step
RHO	relative resistivity of the conductor (copper=1)
RO	roughness of the metalization
Sn	spacing between coupled lines (n is an integer)
SQ	option specifier for square corner bend
SY	option specifier for symmetric step
TA	loss tangent of the dielectric material
TM	strip metalization thickness
UN	units multiplier (mils=0.0254, mm=1.0)
VIAR	radius of a viahole
VIAT	thickness of the metalization in a viahole
W	width of a strip
WA	width of interdigital end-resonator compensating line
WIDE	width on one side of a step (also W)
Whi	width of a high-impedance line
WI	width of an input leader strip
Wlo	width of a low impedance line
WMATCH	width of external matching lines for elliptic bandpass
Wn	width of individual lines (n is an integer)
WOUT	width of an output leader strip
WS	width of the stub side of a tee
WT	width of the through path of a tee
Zhi	impedance of the high-impedance lines in a filter
Zlo	impedance of the low-impedance lines in a filter

B.4 Program Units

The units used in =SuperStar= element codes are

Resistance	ohms
Inductance	nanohenries
Capacitance	picofarads
Conductance	mhos
Frequency	megahertz
Delay	nanoseconds
Length (elec)	degrees
Length (phys)	mils or millimeters

Index